WSP

WASHINGTON SUMMIT PUBLISHERS
2019

Also By Richard Lynn

Dysgenics: Genetic Deterioration in Modern Populations (1996)
Eugenics: A Reassessment (2001)
The Global Bell Curve (2008)
The Chosen People: A Study of Jewish Intelligence and Achievement (2011)
Race Differences in Intelligence (2006 / 2015)

With Tatu Vanhanen

IQ and the Wealth of Nations (2002)
IQ and Global Inequality (2006)
Intelligence: A Unifying Construct for the Social Sciences (2012)

With Edward Dutton

Race and Sport: Evolution and Racial Differences in Sporting Ability (2015)

With David Becker

The Intelligence of Nations (2019)

RACE DIFFERENCES IN PSYCHOPATHIC PERSONALITY

An Evolutionary Analysis

BY RICHARD LYNN

EDITED BY EDWARD DUTTON

Washington Summit Publishers
P.O. Box 100563
Arlington VA 22210

email : hello@WashSummit.com
web: www.WashSummit.com

Cataloging-in-Publication Data is on file with the Library of Congress

  ISBN: 978-1-59368-057-2
eISBN: 978-1-59368-058-9

Printed in the United States of America
10 9 8 7 6 5 4 3 2 1

CONTENTS

FIGURES AND TABLES

RACE
DIFFERENCES
IN
PSYCHOPATHIC
PERSONALITY

INTRODUCTION

The starting point for this inquiry into racial and ethnic differences in psychopathic personality disorder is a problem raised by Richard Herrnstein and Charles Murray in their book *The Bell Curve* (1994), in which they showed that racial and ethnic differences in a number of social problems in the United States, including crime, poverty, long-term unemployment, teenage pregnancy, welfare dependency, and out-of-wedlock births, can be partly explained by differences in intelligence. They reported that Blacks have the lowest intelligence and the highest rates of these phenomena, followed by Hispanics, while Whites have the highest intelligence and the lowest rates of these phenomena. However, they also found that the racial and ethnic differences in these social pathologies cannot be fully explained by differences in intelligence and therefore concluded that some other factor or factors must also be involved. They wrote,

> Some ethnic differences are not washed away by controlling for either intelligence or for any other variables that we examined. We leave those remaining differences unexplained and look forward to learning from our colleagues where the explanations lie. (Herrnstein & Murray, 1994, 340)

In 2002, I took up this invitation and proposed that a component in the solution to this problem lies in racial and ethnic differences in psychopathic personality such that this is highest among Blacks and Native Americans, next highest in Hispanics, lower in Whites and lowest in South Asians and East Asians. This theory attracted little attention and has been ignored in texts on this issue, such as *Violent Crime: Assessing Race and Ethnic Differences*

(Hawkins, 2003), *The Oxford Handbook of Impulse Control Disorders* (Grant & Potenza, 2012), and *Handbook on Psychopathy and Law* (Kiehl & Sinnott-Armstrong, 2013). Hence, in this book, I present a fuller case for the theory that there are racial and ethnic differences in psychopathic personality. We examine this hypothesis for the United States in Chapter 2 and, in subsequent chapters, for Canada, Europe, Africa, Northeast Asia, South Asia, Latin America and the Caribbean, Australia, New Zealand, the Pacific Islands and among the Inuit. We conclude with an examination of the neurology, genetics, and evolution of racial and ethnic differences in psychopathic personality.

1. DEFINITION OF PSYCHOPATHIC PERSONALITY

The condition now known as psychopathic personality was identified in the early 19th century by the French physician Philippe Pinel (1801), who described patients who had "a lack of restraint and whose behavior was marked by a complete remorselessness of their actions" (Perez, 2012, p.519). Three decades later, the British physician John Pritchard (1835) proposed the term "moral imbecility" for those deficient in moral sense but whose intellectual ability was unimpaired. In 1904, the German psychiatrist Emile Kraepelin (1904) introduced the term "psychopathic personality", in his much cited German-language book, to describe the condition, and this has been employed as a diagnostic label throughout the 20th century and into the 21st. In 1941, the condition was described by Hervey Milton Cleckley (1941, 1976) in his classic book *The Mask of Sanity*. He described the criteria for the condition as being a "general poverty of affect," defective insight, absence of nervousness and anxiety, lack of remorse or shame, superficial charm, pathological lying, egocentricity, inability to love, failure to establish close or intimate relationships, irresponsibility, impulsive psychopathic acts, failure to learn from experience, reckless behavior under the influence of alcohol, and a lack of long-term goals.

In 1984, the American Psychiatric Association dropped the term "psychopathic personality" and replaced it with "antisocial personality disorder." Some authorities, such as Lykken (1995), regard this as simply a synonym for psychopathic personality. Others, such as Hare (1994), consider that there is some difference between the two concepts and that antisocial personality disorder is a less satisfactory term, because it fails to give sufficient emphasis to the psychological features as opposed to the behavioral characteristics of the condition. Despite these fine distinctions, for practical purposes antisocial personality disorder and psychopathic personality can be regarded as largely synonymous descriptions of the same condition and are treated as such in this book. The term sociopathy has been used by Lykken (1995) to describe psychopathic behavior that has largely environmental origins in poor socialization, while psychopathy has a largely genetic origin; that said, behaviorally sociopathy and psychopathy are indistinguishable and are treated as synonymous with psychopathic personality. The term "borderline personality disorder" is also sometimes used to designate a milder form of the condition or sometimes to designate a condition with high emotional reactivity in contrast to the low emotional reactivity typical of psychopathic personality (Chabrol, Valls, van Leeuwen & Bui, 2012).

In 1994 the American Psychiatric Association (1994) issued a revised Diagnostic Manual in which it listed 11 features of psychopathic personality disorder.

These are:

1. inability to sustain consistent work behavior;

2. failure to conform to social norms with respect to lawful behavior;

3. irritability and aggressivity, as indicated by frequent physical fights and assaults;

4. repeated failure to honor financial obligations;

5. failure to plan ahead or impulsivity;

6. no regard for truth, as indicated by repeated lying, use of aliases, or "conning" others (an association between psychopathy and self-reported lying has been reported by Giammarco, Atkinson, Baughman, Veselka & Vernon, 2013);

7. recklessness regarding one's own or others' personal safety, as indicated by driving while intoxicated or recurrent speeding;

8. inability to function as a responsible parent;

9. failure to sustain a monogamous relationship for more than one year;

10. lacking remorse; and

11. the presence of conduct disorder in childhood.

This is a useful list of the principal constituents of the condition, subject to the reservation that it does not explicitly include the deficiency of moral sense, although this is implicit in virtually all the listed behaviors. The American Psychiatric Association (2000) issued a further revision dividing psychopathic personality disorder into three clusters designated (A) odd or eccentric; (B) dramatic, emotional or erratic; and (C) anxious or fearful. It also proposed the concept of borderline personality disorder as a milder form of antisocial personality disorder, which is characterized by "a pervasive pattern of instability in emotional regulation, impulse control, interpersonal relationships and self-image." Those with borderline personality disorder are weak on emotional intelligence (Sinclair & Feigenbaum, 2012). Despite the replacement of the concept of psychopathic personality with that of psychopathic personality disorder by the American Psychiatric Association, many psychiatrists and psychologists regard the term psychopathic personality as preferable and continue to use it.

Psychopathic personality is not a discrete condition, but the extreme expression of a personality trait that is present in greater or lesser degrees throughout the population. Thus,

> despite the media's portrayal and the general public's conception of the psychopath as seemingly inhuman and fundamentally unlike most people, the empirical evidence from large-scale studies suggests that psychopathic traits are continuously distributed, present in samples from the community and the corporate world, and linked to

common genetic factors. (Neumann, Schmitt, Carter, Embley & Hare, 2012)

Psychopathic personality/anti-social personality disorder is most commonly measured by Hare's Psychopathy Checklist, a 20-item rating scale of expressions of psychopathic personality that gives a score for the strength of the condition (Hare, 2003). Community studies have found that psychopathic traits are associated with high levels of drug and substance abuse, alcohol use and violence (Neumann & Hare, 2008), criminal offenses and other externalizing psychopathology (Neumann & Pardini, 2014), and to morally problematic corporate behavior (Babiak et al., 2010).

2. SUBFACTORS OF PSYCHOPATHIC PERSONALITY

Psychopathic personality is a general factor that has been divided into two or more subfactors. In the analyses by Robert Hare and his colleagues, psychopathic personality has been found to contain two correlated subfactors. Factor 1 is designated "emotional-interpersonal" and consists of glibness, superficial charm, grandiose sense of self-worth, pathological lying, conning and manipulative behavior, lack of remorse or guilt, shallow affect, callousness, lack of empathy, failure to accept responsibility for actions, and the selfish and remorseless use of others. Factor 2 is designated "social deviance" and consists of a syndrome of socially deviant and psychopathic behaviors comprising the need for stimulation, proneness to boredom, parasitic lifestyle, poor control of behavior, early behavior problems, lack of realistic long-term goals, impulsivity, irresponsibility, and juvenile delinquency (Harpur, Hare & Hakstian, 1989; Hare, 1991, 1994; Harpur, Hart & Hare, 1994). These two factors are found in factor analyses of the 20 item Psychopathic Checklist, which also contains promiscuous sexual behavior and criminality that are associated about equally with both of these factors. The two factors are positively correlated at 0.39, showing the presence of a single general factor (Akhtar, Ahmetoglu

& Chamarro-Premuzic, 2013). The same two positively correlated factors have been identified in Poland by Debowska, Boduszek, Kola, and Hyland (2014).

Cooke and Michie (2001) have proposed a three-factor model of psychopathic personality that splits the items associated with Factor 1 into affective ("affectively deficient") and interpersonal ("arrogant and deceitful") factors and have eliminated several items associated with Factor 2, based on findings that they were poor indicators of psychopathic personality, to create a new behavioral factor ("impulsive and irresponsible lifestyle"). They claim that the three-factor model fits data better on correctional and psychiatric samples than does the two-factor model. In a more recent analysis, Hare (2003) has proposed a four-factor model of psychopathy, consisting of interpersonal, affective, lifestyle, and psychopathic traits; further support for this model has been presented by Neumann (2007) and Hare and Neumann (2008). However, Hare (2003) has noted that his four factors are correlated and "provide a viable representation of the larger psychopathy construct."

3. CONDUCT DISORDER

The concept of psychopathic personality is normally not used for children or young adolescents up to the age of around 15 years. Children and young adolescents manifesting psychopathic behaviors are instead identified as having conduct disorders. The principal criteria set out by the American Psychiatric Association (1994) for a diagnosis of conduct disorder are persistent stealing, lying, truancy, running away from home, fighting, bullying, arson, burglary, vandalism, sexual precocity, and cruelty. Childhood conduct disorder is therefore similar to psychopathic personality in older adolescents and adults. A diagnosis of psychopathic personality is not normally made without evidence of conduct disorder in childhood. A number of studies have shown that conduct disorder in children is a frequent precursor of psychopathic personality in later adolescence and adulthood (Bernstein, Cohen, Skodal, Bezirganian & Brook, 1996; Loeber, 1990; Mealy, 1995).

It is estimated that conduct disorder has a prevalence of nine percent in young males under the age of 18 in the United States (Farrington, 1991). Conduct disorder in children is more common than psychopathic personality, and thus not all children with conduct disorder mature into psychopathic personalities. A study in Britain has reported that 40 percent of boys and 35 percent of girls with conduct disorder developed antisocial personality disorder as adults (Zoccolillo, Pickles, Quinton & Rutter, 1992). Conduct disorder is most generally assessed by rating scales of which the most widely used are the Child Behavior Checklist (CBCL) constructed by Achenbach (1992) and the Teachers Rating Scale (TRS) constructed by Conners (1989). These rating scales consist of the number of expressions of conduct disorder that the teacher, some other professional, or the parent has identified as manifested by the child being assessed.

4. PREVALENCE OF PSYCHOPATHIC PERSONALITY

The prevalence of psychopathic personality varies with age. As noted above, it is estimated that conduct disorder, the precursor of psychopathic personality in children, has a prevalence of 9 percent in young males under the age of 18 in the United States. Bernstein, Cohen, Velez, et al. (1993) carried out a study of the prevalence of personality disorders over a two-year period in adolescents aged from 9 to 19 years. They reported that the prevalence of personality disorders peaked at age 12 in boys and at age 13 in girls and declined thereafter. Moffit (1993) has reported that conduct disorder and psychopathic personality increase from the age of 7 to 17 years and then decline from the age of 17 to 50 years to around 2 to 3.5 percent. A review of a number of studies of the prevalence of psychopathic personality disorders among older adults in the United States by Torgersen, Kringlen, and Cramer (2001) concluded that in the White population, there is a rate of 0.7 percent and a further 0.7 percent of borderline personality disorder. A more recent review by Cartwright (2012) concluded that borderline personality disorder has a prevalence rate of approximately two percent in the general population.

5. SEX DIFFERENCES IN PSYCHOPATHIC PERSONALITY

It has invariably been found that more males than females are psychopathic. Studies of lifetime prevalence rates showing this in a number of countries are summarized in Table 1.1. A study of borderline personality reports a male-female ratio of 3:1 (Compton et al., 2005).

Table 1.1. Sex differences in psychopathic personality (percentages)

COUNTRY	MALES	FEMALES	REFERENCE
Canada	6.5	0.8	Bland et al., 1988
Hong Kong	2.8	m0.5	Chen et al., 1993
New Zealand	4.2	0.5	Wells et al., 1989
South Korea	3.5	0.8	Lee et al., 1990
USA	4.5	0.8	Robins et al., 1991
USA	6.2	2.4	Samuels et al. 2002

Studies showing that males obtain higher scores than females on measures of psychopathic characteristics among children and adolescents are summarized in Table 1.2. Row 1 gives results for externalizing behavior for 5-18 year olds in Hawaii. Rows 2 and 3 give results for callous-unemotional and psychopathic behaviors from a study of the Youth Psychopathic Traits Inventory in a sample of 972 French 17-year-old school students.

Table 1.2. Sex differences in psychopathic personality (ds)

AGE	TRAIT	MALES	FEMALES	*d*	REFERENCES
5-18	Externalizing behavior	15.1 (16.4)	9.0 (12.4)	.42	Loo & Rapport, 1998
17	Callous-unemotional	32.7 (5.6)	30.3 (5.0)	.45	Chabrol et al, 2012
17	Psychopathic behavior	12.0 (10.7)	6.3 (6.5)	.66	Chabrol et al, 2012

6. PSYCHOPATHIC PERSONALITY AND INTELLIGENCE

Evidence is mixed on the relationship between psychopathic personality and intelligence. Harpur, Hare, and Hakstian (1989) obtained a near zero correlation between Factor 1 of the Psychopathy Checklist–Revised (PCL–R) and IQ (r =.04) and a negligible negative correlation between Factor 2 and IQ (r =-.15). Others who have reported no significant relationship between intelligence and psychopathy include Dahlstrom, et al. (1986, p. 243), Hart, Forth, and Hare (1990), Gladden, Figueredo, and Jacobs (2008) and Allen, et al. (2013). On the other hand, it has been known for some decades that criminals have below average intelligence. Thirty years ago, Wilson and Herrnstein (1985, p. 159) wrote,

> for four decades large bodies of evidence have consistently shown about a ten-point IQ gap between the average offender and nonoffender in Great Britain and in the United States.

More recently, Clarizio (1997) has summarized several studies reporting below average IQs in children with conduct disorders, delinquents, and adult psychopaths.

Evidently, psychopathic personality is a robust concept and high psychopathic personality is associated with many kinds of anti-social and dangerous behavior. If there were race differences in its prevalence, this would be significant reverberations throughout multicultural societies, and it would have profound consequences. In this book, we will see that there are, indeed, clear and consistent race differences in psychopathic personality.

UNITED STATES

1. PREVALENCE OF PSYCHOPATHIC PERSONALITY

Studies of race differences in the prevalence of psychopathic personality in the U.S. population are summarized in Table 2.1.

Table 2.1. Race differences in the prevalence of psychopathic personality (percentages)

	ASIAN	BLACK	HISP.	NATIVE AMERICAN	WHITE	REFERENCE
1	-	3.7	-	-	2.7	Malzberg, 1944
2	-	11.0	-	-	3.0	Tracy et al., 1990
3	-	.29	-	-	.00	Morey, 1991
4	-	25.6	-	-	18.1	Compton et al., 2000
5	10.1	16.6	14.0	24.1	14.6	Huang et al., 2006

Row 1 gives the results of an early study of admissions to all psychiatric institutions in New York state in 1929-1931, showing higher percentages of Blacks than Whites diagnosed with psychopathic personality. Row 2 gives the results of a study of a

1958 birth cohort of boys in Philadelphia, giving the percentages that became chronic recidivists as 3.7 times greater among Blacks than among Whites. Row 3 gives the results of a study of the Psychopathic Features scale of the Personality Assessment Inventory on a normative sample of 851 White and 117 Black community residents from 12 states. Blacks obtained a higher score than Whites of 2.9 T-score points, i.e., an effect size difference of .29. These data were obtained from self-reports of psychopathic behavior, which likely underestimate the true difference. Row 4 gives the results of psychopathic personality in a sample with drug dependence, showing a higher rate of disorder among Blacks than Whites. Row 5 gives the results of a large community study of personality disorders, which includes drug addiction and psychopathic disorders, and shows the highest rate among Native Americans, followed by Blacks, Whites, and Hispanics, and the lowest among Asians.

These results are inconsistent with two epidemiological studies that reported no significant differences in the percentages of Blacks, Hispanics, and Whites with lifetime prevalence of psychopathic personality. These are the Environmental Catchment Area Study (Robins & Regier, 1991) and the National Comorbidity Study (Kessler, McGonagle, Zao et al, 1994). The most probable explanation for these anomalous results is that they were obtained from interviewers who asked respondents whether they had ever committed a number of criminal and psychopathic acts. In these interviews, it is likely that many respondents did not disclose the full extent of their criminal and psychopathic behavior. This is particularly the case with psychopaths, for whom "no regard for the truth" is a central characteristic. It has been found in two studies that Blacks have approximately the same rate of self-reported crime as Whites, although records showed that their crime rates are considerably higher (Hindelang, Hirshi & Weis, 1981). The first of these was the Richmond Youth Project, in which Black and White youths were interviewed and asked about their criminal activities. The Black-White ratio for self-reported criminal offenses was a negligible 1.1: 1.0, but the actual ratio obtained from police records was 2.0: 1.0. In the second study (the Seattle Project), the Black-White ratio for self-reported criminal offenses was 0.95:1.0, while the police records showed that the actual ratio was 1.6: 1.0. The authors conclude that "blacks failed to report known official offenses at a much higher rate than whites." (p. 180) The same conclusion has been reached by Huizinger & Elliott (1984). The lesson to be drawn from these studies is that race differences in self-reported psychopathic behavior need to be treated with skepticism and more reliance placed on objective behavior such as rates of crime, exclusions from school, unstable marital relationships and many others documented later in this and subsequent chapters.

2. THE MMPI PSYCHOPATHIC DEVIATE SCALE

A number of studies of race differences in the prevalence of psychopathic personality have been carried out using the Minnesota Multiphasic Personality Inventory (MMPI). This questionnaire was constructed in the late 1930s by Hathaway and McKinley (1940), and is the most widely used instrument for the assessment of abnormal personality in the world (Pace et al., 2006). It consists of a series of scales for the measurement of psychiatric conditions regarded as continuously distributed in the population, such as hysteria, mania, depression and psychopathic personality. The scale for the measurement of psychopathic personality is the Psychopathic Deviate Scale. This was constructed by writing a number of questions, giving them to criterion groups of those manifesting psychopathic behavior and "normals," and selecting for the scale the questions best differentiating the two groups. The criterion group manifesting psychopathic behavior consisted of 17–24 year olds appearing before the courts and referred for psychiatric examination because of their "long histories of delinquent type behaviors such as stealing, lying, alcohol abuse, promiscuity, forgery and truancy" (Archer, 1997, p. 20). The common feature of this group has been described as their failure to "learn those anticipatory anxieties which operate to deter most people from committing psychopathic behavior." (Marks, Seeman & Haller, 1974, p. 25) The manual describes those scoring high on the scale as irresponsible, psychopathic, aggressive, having recurrent marital and work problems, and underachieving (Hathaway & McKinley, 1989).

A number of studies have shown that the Psychopathic Deviate scale differentiates delinquents and criminals from non-delinquents and non-criminals (e.g. Elion & Megargee,1975).

During the seventy or so years following its publication, the MMPI has been administered to numerous groups. Normative data for the means obtained in the United States by "normal" (i.e. non-psychiatric) samples of Asians, Blacks, Whites, Hispanics, and Native Americans on the Psychopathic Deviate scale of the MMPI are summarized in Table 2.2. These are calculated as *d* scores (differences expressed in standard deviation units) in relation to a White value of zero, positive signs indicating scores higher than those of Whites, and negative signs indicating scores lower than those of Whites.

Table 2.2. Race differences (d) on the Psychopathic deviate scale of the MMP1 and MMPI-2

	ASIAN	BLACK	HISPANIC	NATIVE AMERICAN	WHITE	REFERENCE
1	-.31	.29	.00	.44	.00	Dahlstrom et al., 1986
2	-.18	.48	.70	.74	.00	Hathaway & McKinley, 1989
3	-	.33	-	-	.00	Timbruck & Graham, 1994
4	-	.33	-	-	.00	Timbruck & Graham, 1994
5	-	.32	-	-	.00	Goldman et al., 1995
6	-	.17	-	-	.00	Goldman et al., 1995

	ASIAN	BLACK	HISPANIC	NATIVE AMERICAN	WHITE	REFERENCE
7	-	.33	.36	-	.00	Archer, 1997
8	-	-	.16	-	.00	Nagy et al., 1997
9	-	.35	-	-	.00	Arbisi et al., 2002
10	-	.32	-	-	.00	Arbisi et al., 2002
11	-	-	-	.57	.00	Robin et al., 2003
12	-	-	-	.53	.00	Pace et al., 2006

Row 1 shows that Blacks and Native Americans obtain higher means than Whites on the scale by 0.29d and 0.44d, respectively. Hispanics obtain the same mean as Whites, while Asians obtain a lower mean (-0.31d). Row 2 gives results from the standardization sample of the MMPI-2, a revision of the test standardized in the 1980s (Hathaway & McKinley, 1989). The standardization sample numbered 2,500, and was selected to match the national population of the United States in terms of geographical location, age, educational level, socioeconomic status, earnings, marital status, and ethnicity. Of the 54 questions in the Psychopathic Deviate scale of the original test, four were replaced in the MMPI-2. The manual provides means and standard deviations for Whites, Blacks, Asians, Native Americans, and Hispanics. No details are given regarding the national or ethnic origins of the Asian group. The means of the five groups have been converted to d scores in relation to a White zero and combined males and females, in the same way as for the original MMPI. The results are consistent with those of the original MMPI in so far as Blacks and Native Americans obtained higher mean scores than Whites, while Asians obtained lower scores. The result for Hispanics is inconsistent in so far as they obtained a substantially higher mean than Whites in the MMPI-2 standardization sample, but the same mean as Whites in the original MMPI. Rows 3 and 4 give results from a sample of 292 male and female Blacks and 292 male and female Whites for the MMPI-2, showing that in both sexes Blacks had higher scores by .33d. Row 5 gives results for 48 Black and 68 White male college students for the MMPI-2, showing that Blacks had a higher score by .32d. Row 6 gives results for 58 Black and 64 White female college students for the MMPI-2, showing that Blacks had a higher score by .17d. Row 7 gives results for the MMPI-A published in 1992 as a version of the test designed for adolescents. Normative data are given for Blacks, Whites, and "others", entered as Hispanics because these are by far the largest "other" ethnic group. It will be seen that Blacks and Hispanics score about a third of a standard deviation higher than Whites. Row 8 gives results from the MMPI-A for Hispanics and Whites, again showing that Hispanics score higher than Whites. Rows 9 and 10 give results from a sample of 159 Black and 1,233 White male and female psychiatric patients for the MMPI-2, showing that male and female Blacks had higher scores than Whites by .35d and .32d, respectively. Row 11 gives results for the MMPI-2 for 832 Native Americans showing that they score higher than Whites by .57d. Row 12 gives results from a further sample (n=171) for the MMPI-2 for Native Americans, confirming that they score higher than Whites by .57d. Zuckerman (2003, p.1465) has objected to the data on the race differences shown in Table 2.2, on the grounds that, "It is only when the Psychopathic deviate scale is high and accompanied by a peak on the Hypomania scale that the results can be interpreted as indicative of a psychopathic personality". To examine this objection, race differences on the Hypomania scale of the MMPI and MMPI2 are shown in Table 2.3. It will be seen that Blacks, Hispanics and Native Americans, all score considerably higher than Whites, while in three of the four entries East Asians score a little lower than Whites.

Table 2.3. Race differences (d) on the Hypomania scale of the MMP1 and MMPI-2

TEST	SEX	BLACK	EAST ASIAN	HISPANIC	NAT AM.	WHITE	REFERENCE
MMPI	M	.47	.13	.62	-	.00	Dahlstrom et al., 1986
MMPI		.39	-.16	.62	-	.00	Dahlstrom et al., 1986
MMPI-2	M	.40	-.14	.28	.47	.00	Hathaway & McKinley, 1989
MMPI-2	F	.50	-.07	.44	.95	.00	Hathaway & McKinley, 1989
MMPI-2	M	.36	-	-	-	.00	Timbruck & Graham, 1994
MMPI-2	F	.31	-	-	-	.00	Timbruck & Graham,1994
MMPI-2	M	.57	-	-	-	.00	Arbisi et al.,2002
MMPI-2	F	.59	-	-	-	.00	Arbisi et al., 2002
MMPI-2	M/F	-	-	-	.56	.00	Robin et al., 200

3. CONDUCT DISORDERS

Studies reporting race differences in conduct disorders are summarized in Table 2.4. The results of the first five studies have been calculated as *d* scores (standard deviation units) in relation to Whites set at zero. Row 1 presents data for Blacks and Whites from a study of 1,027 children in North Carolina assessed for conduct disorders by teachers. The data shows mean scores about half a standard deviation higher in Blacks as compared with Whites. Rows 2 to 4 present further data from various locations in the continental United States in which conduct disorders were assessed by teachers. Row 5 gives results for oppositional defiant disorder assessed by teachers, showing that this is higher in Blacks and Hispanics than in Whites. Row 6 gives results from a study of disruptive disorder among 18 year olds expressed as an odds ratio, showing a higher rate in Blacks than in Whites.

Table 2.4. Race differences in conduct disorders (d; OR)

ASIAN	BLACK	HISPANIC	NAT. AMER.	WHITE	REFERENCE
-	.53	-	-	.00	Epstein et al., 1998
-1.12	.49	-	-	.00	Feng & Cartledge,1996
-	-	-	.35	.00	Dion et al., 1998
-.56	-	-	-	.00	Chang et al.,1995
-	.59	.28	-	.00	Arnold et al., 2003
-	1.48	-	-	1.0	Cuffe et al, 2005

A number of studies have reported race differences in conduct disorders as percentages of populations, and these are summarized in Table 2.5. Row 1 shows the percentage of conduct disorder to be lowest in Asians, a little higher in Whites, and substantially higher in Blacks and Hispanics. Row 2 shows a high percentage of conduct disorder in Native Americans. Row 3 gives the results of a study of 4,347 16 year olds in California and Oregon who reported on their own "deviant behaviors," consisting of cheating on a test, truancy, being sent out of class, and stealing. The results show greater deviance among Blacks than among Whites and Asians. The high percentages for all groups indicate that the criterion for deviance was more relaxed than for conduct disorders than in other studies. Rows 4 and 5 give data from a 2009 study of 18,819 teenage school students who reported whether they had truanted from school for one to three days or four plus days during the last month. The data shows the highest rate in Hispanics followed by Blacks and the lowest rate in Whites. Row 6 gives the results of a national study of 21,260 6, 8 and 10 year olds and shows the percentage of conduct disorder to be lowest in Asians, highest in Blacks, and intermediate in Hispanics and Whites. Row 7 gives the results of a study of race differences in conduct disorders as odds ratios (ORs) with the percentages of conduct disorders of Whites set at 1.0, and shows that Black children had 1.6 times the conduct disorder rate of Whites, while Hispanics had the same rate as Whites.

Table 2.5. Race differences in conduct disorders (ORs)

	ASIAN	BLACK	HISP.	NAT. AMER.	WHITE	REFERENCE
1	3.0	5.9	8.0	-	3.4	Miller et al., 1995
2	-	-	-	17	-	Kunitz at al., 1995
3	37	51	52	-	39	Ellikson & Morton, 1999
4	-	9.1	12.2	-	8.7	Vaughn et al., 2013

	ASIAN	BLACK	HISP.	NAT. AMER.	WHITE	REFERENCE
5	-	1.8	2.7	-	1.4	Vaughn et al., 2013
6	0.8	1.9	1.6	-	1.6	Bates, 2013
7	-	1.6	1.0	-	1.0	McDermott & Spencer,1997

The prevalence of conduct disorder is also expressed in the extent to which youths get into fights. Studies of this are summarized in Table 2.6, and they show the highest percentages in Blacks and the lowest in Whites, with Hispanics intermediate. Row 3 gives the results of self-reported violent attacks in the NLSY 1997 sample (average age 17) showing a significantly higher rate in Blacks.

Table 2.6. Race differences in conduct disorders in children (percentages)

	BLACK	CUBAN	HISPANIC	PUERTO RICAN	WHITE	REFERENCE
1	45	-	40	-	32	CDC, 2008
2	54	49	50	48	38	Estrada-Martínez et al., 2013
3	.55	-	-	-	.38	McNulty et al., 2013

A further expression of conduct disorder is juvenile delinquency. Studies of this are summarized in Table 2.7. The results are presented as odds ratios giving the numbers of Blacks to one White. Rows 1 and 2 show that in the 1960s the ratio of Blacks to Whites for criminal convictions for boys was 8.1:1, while for girls the ratio of Blacks to Whites was 14.1:1. Rows 3 and 4 show similar ratios for institutionalization for criminal offenses. Row 5 gives the results of studies showing a Black-White ratio of 4.8:1 for police contacts arising from delinquent behavior. Row 6 gives the results of studies showing a Black-White ratio of 3.9:1 for imprisonment rates for juvenile crime for 1997-2006 institutions.

Table 2.7. Race differences in delinquency as odds rations

	OFFENCE	SEX	BLACK	WHITE	REFERENCE
1	Conviction	M	8.1	1.0	Gold,1966
2	Conviction	F	14.1	1.0	Gold,1966
3	Institutionalization	M	9.8	1.0	Gold,1966
4	Institutionalization	F	13.9	1.0	Gold,1966
5	5+ Police contacts	M	4.8	1.0	Hindelang et al., 1981
6	Imprisonment	MF	3.9	1.0	Davis & Sorensen, 2013

4. SCHOOL SUSPENSIONS AND EXCLUSIONS

Children with conduct disorders are sometimes suspended or expelled from schools because of their constant misbehavior. The principal reasons for suspensions and expulsions are "disobedience in various forms - constantly refusing to comply with school rules, verbal abuse or insolence to teachers." (Gillborn & Gipps, 1996, p. 53) Racial and ethnic differences in suspensions and exclusions as further measures of conduct disorders are given in Tables 2.8 and 2.9.

Table 2.8 gives race differences in school suspensions and expulsions expressed as percentages of school populations.

Table 2.8. Race differences in school suspensions and exclusions (percentages)

	ASIAN	BLACK	HISP.	NAT. AMER.	WHITE	REFERENCE
1	-	32.1	-	-	12.8	Backman, 1970
2	-	22.2	-	-	7.6	Elliott et al., 1980
3	-	12.8	-	-	4.1	Children's Defence Fund, 1985
4	-	19.6	-	-	3.7	Munsch & Wampler, 1993
5	3.2	12.8	9.5	11.0	8.4	Gordon et al., 2000
6	3.0	20.0	9.0	12.0	7.0	US Dept Education, 2011
7	1.0	11.0	4.0	6.0	3.0	US Dept Education, 2011
8	-	31	-	-	11	Wright et al., 2014

In the first of these studies, Backman (1970) reported that Blacks were 2.5 times more suspended and excluded than Whites. Rows 2, 3 and 4 give three further studies confirming this result. Row 5 gives the largest of these studies reporting the results collected in 1999 for 1.8 million school children. In this instance, data was drawn from public schools in Chicago, San Francisco, Durham NC, Denver, Austin TX, Boston, Los Angeles, Miami, Missoula MT, Providence RI, and Salem OR. Results for the five major racial and ethnic groups showed the greatest suspension and exclusion rate for Blacks (12.8 percent), followed in descending order by Native Americans (11.0 percent), Hispanics (9.5 percent), Whites (8.4 percent), and East Asians (3.2 percent). Rows 6 and 7 give suspension rates in 2009-2010 of K-12 students (18 year olds) for boys and girls nationwide. Row 8 shows results for the percentages of Blacks and Whites aged 4 through 15 who had ever been suspended from school, and show the percentage of Blacks to be 3.8 times higher than that of Whites. The authors show that these differences are explained by differences in problem behaviors and not by teacher discrimination. Taken as a whole, the studies show the highest rates of school suspensions and exclusions are among Blacks, followed by Native Americans, Whites, and the lowest rates are among Asians.

Race differences in school suspensions and exclusions, expressed as odds ratios in which the rate for Whites is set at 1.0, are given in Table 2.9.

Table 2.9. Race differences in school suspensions and exclusions as odds ratios

BLACK	HISPANIC	NATIVE AMERICAN	WHITE	REFERENCE
3.50	-	1.52	1.0	Bickel & Qualls, 1980
2.84	1.23	1.98	1.0	APA, 2008
2.47	1.50	-	1.0	APA, 2008

Row 1 shows that Blacks have 3.5 times the rate of school suspension and exclusion than Whites, confirming results given in Table 2.8. Rows 2 and 3 give results from a Task Force set up in 2004 by the American Psychological Association (APA) to examine these racial differences in school suspensions and exclusions, focusing on whether "zero tolerance" attempts to reduce these had been successful, and also to consider the explanation for the race differences. The Task Force reported its conclusions in 2008 (APA Zero Tolerance Task Force, 2008). It documented new evidence for race differences in school suspensions and exclusions for the school year 2002-2003, and these are given in rows 2 and 3 of Table 2.9. Row 2 gives results for suspensions and shows that Blacks had the highest rate at 2.84 times that of Whites, Native Americans were next at 1.52 times, while Hispanics were 1.23 times more suspended than Whites. Row

3 gives results for exclusions and shows that Blacks had the highest rate at 2.47 times that of Whites, while Hispanics were 1.50 times more than Whites. These studies provide further confirmation of the highest rates of school suspensions and exclusions for Blacks, followed by Native Americans and Hispanics, and the lowest rates for Whites.

The APA Task Force report considered the reason for these race differences and concluded that "there are no data supporting the assumption that African American students exhibit higher rates of disruption or violence that would warrant higher rates of discipline. Rather, African American students may be disciplined more severely for less serious or more subjective reasons...the disproportionate discipline of students of color may be due to lack of teacher preparation in classroom management, lack of training in culturally competent practices, or racial stereotypes." (p. 854) This is a remarkable assertion because the most common reason for school suspensions and exclusions is conduct disorder (also termed behavior problems or "oppositional defiance disorder"), consisting of excessive aggression, violence, disobedience, and criminal offenses such as drug dealing, and a number of studies summarized in this chapter have reported that racial differences in these are similar to those in suspensions and exclusions. The most straightforward explanation for the race differences in school suspensions and exclusions is that these are a result of the differences in levels of conduct disorder.

A further curious feature of the Task Force's report is that is that it failed to mention that fewer Asians than Whites are suspended and excluded from schools. This undermines the Task Force's suggestion that teachers' racial stereotypes are responsible for the high rate of suspensions and exclusions of Hispanics, Native Americans, and African Americans, unless the Task Force wished to suggest that teachers are more prejudiced against Whites than against Asians. There is nothing surprising about the low rate of school suspensions and exclusions of Asians. This is to be expected because of the studies showing that Asians have a lower rates of conduct disorder than Whites and other groups.

We are left with the puzzle of why the Task Force failed to mention that race differences in school suspensions and exclusions are consistent with those in conduct disorders, and in a number of other expressions of psychopathic behavior reviewed throughout this book. We are similarly also left with the puzzle of why the Task Force failed to mention the low rate of school suspensions and exclusions of Asians. Perhaps the Task Force was unaware of these studies, but this seems improbable. More likely, the Task Force was aware of them but preferred to ignore them and blame the higher rates of suspensions and exclusions of Hispanics, Native Americans, and African Americans on the "lack of teacher preparation in classroom management, lack of training in culturally competent practices, or racial stereotypes" of white teachers." Whatever the explanation, the conclusion of the American Psychological Association Task Force that "there are no data supporting the assumption that African American students exhibit higher rates of disruption or violence that would warrant higher rates of discipline," (p. 854) can only be regarded as inaccurate.

Furthermore, the rate of school expulsions and suspension for Blacks reported by the Task Force is only 2.6 times that of Whites, while the Black rates for delinquency and criminal convictions shown in Table 2.7 are considerably greater than these. This suggests that teacher bias is unlikely to be the reason for the racial difference in school expulsions and suspensions. To the contrary, the greater differences in criminal convictions suggests that teachers are more tolerant of psychopathic behavior in Blacks, and are reluctant to suspend and exclude them, probably for fear of being accused of racism.

All the evidence for race differences in psychopathic personality shows that "lack of teacher preparation in classroom management, lack of training in culturally competent practices, or racial stereotypes held by teachers," cannot be the sole explanation for the greater proportion of Blacks that are suspended and excluded from school. The most reasonable interpretation of all these studies is that there are racial differences in psychopathic behavior, and these provide the most straightforward explanation for the racial differences in school suspensions and exclusions.

5. ATTENTION DEFICIT HYPERACTIVITY DISORDER

Attention Deficit Hyperactivity Disorder (ADHD) consists of the inability to pay attention, restlessness, distractibility, and hyperactivity, and is a prominent feature of children with conduct disorder. In a review of research on the condition, Barkley (1997, p. 65) writes that,

> ADHD is associated with greater risks for low academic achievement, poor school performance, retention in grade, school suspensions and expulsions, poor peer and family relations, anxiety and depression, aggression, conduct problems and delinquency, early substance experimentation and abuse, driving accidents and speeding violations, as well as difficulties in adult social relationships, marriage and employment.

ADHD is therefore a frequent expression of personality disorder. ADHD in childhood persists into adolescence in around 50–80 percent of cases, and into adulthood in about 30-50 percent of cases, where it is associated with personality disorder and crime (Barkley, 1997; Vitelli, 1996; Gordon, Donnelly & Williams, 2014). The association between ADHD and conduct disorder in childhood, and psychopathic personality and crime in adulthood, leads to the expectation that there should be racial differences in ADHD. Studies of race differences in ADHD expressed as ds in relation to Whites set at zero are summarized in Table 2.10. Positive signs denote higher rates of ADHD, and are consistently present in Blacks. Negative signs denote lower rates of ADHD and are consistently present in East Asians. Row 1 gives differences in restlessness and distractibility as assessed by teachers in a sample of 1,337 fourth graders, and shows high scores among Blacks and low scores among East Asians, with Whites and Hispanics intermediate. Row 2 gives data for ADHD assessed by a teacher rating scale for 2,000 children and adolescents for a representative sample for the United States, and shows scores for ADHD about half a standard deviation higher among Blacks than among Whites. Row 3 gives data from another study confirming this. Row 4 provides further confirmation in a study carried out in a large American mid-western city and also gives information for East Asians, who score almost a full standard deviation lower than Whites. Row 5 confirms the low scores of East Asian children from a study carried out in New York. Row 6 gives further results for ADHD assessed by teachers showing higher rates in Blacks and Hispanics than in Whites.

Table 2.10. Race and ethnic differences in attention deficit hyperactivity disorder (ds)

	EAST ASIAN	BLACK	HISPANIC	WHITE	REFERENCE
1	-.71	.59	-.01	.00	Spring et al.,1977
2	-	.46	.15	.00	Du Paul et al., 1997
3	-	.42	-	.00	Epstein et al., 1998
4	-.89	.59	-	.00	Feng & Cartledge, 1996
5	-.35	-	-	.00	Chang et al., 1995
6	-	.31	.27	.00	Arnold et al., 2003

Table 2.11 gives two further studies of race differences in ADHD. Rows 1 and 2 show the percentage of ADHD in Blacks about double that of Whites. The studies summarized in Tables 2.10 and 2.11 show higher rates of ADHD in Blacks than in Whites and lower rates in East Asians, mainly ethnic Chinese. The results for Hispanics are inconsistent, the results in row 1 of Table 2.10 showing the same prevalence as among Whites. Rows 2 and 6 of Table 2.10 and row 2 of Table 2.11 show a higher prevalence than for Whites, and row 1 of Table 2.11 shows a lower prevalence than for Whites.

Table 2.11. Race differences in attention deficit hyperactivity disorder (percentages)

	BLACK	HISPANIC	WHITE	REFERENCE
1	29.6	7.2	15.2	Langsdorf et al., 1979
2	17	-	8	Andretta et al., 2013

6. MORAL UNDERSTANDING

Weakness of moral understanding is a central feature of psychopathic personality. A theory of the development of moral understanding, and a test to measure it, has been formulated by Kohlberg (1976). Over two dozen studies have found that delinquents perform poorly on this test (Raine, 1993). A similar test (the Defining Issues Test) has been developed by Rest (1979, 1986). This consists of a number of stories in which the leading actor is confronted with a moral dilemma. The problem is to discern the moral principle involved, and choose the proper course of action consistent with it. Rest (1979, p. 107) gives a mean score of 21.90 (Sd 8.5) for junior high school students in the American standardization sample of 1,322. A mean score of 18.45 is given for a sample of Black junior high school students by Preston (1979), cited by Rest (1979, p. 64), showing weaker moral understanding by $0.41d$ consistent with greater psychopathic personality.

This result has been confirmed in a study by Sampson and Bartusch (1998). In a sample of 8,782, higher proportions of Blacks showed greater weakness of moral understanding by greater endorsement of the statements such as "laws are made to be broken", "there are no right or wrong ways to make money," and "it is okay for a teenager to have fist fights." The authors and Skeem et al. (2003, p.1451) argue that the race difference is "effectively eliminated after statistically controlling for the extent to which residents' neighborhoods were disadvantaged." This suggests that neighborhood context is as plausible a basis for "normative orientation towards law and deviance as a racially induced subcultural system." However, as neighborhood disadvantage is strongly correlated with race at 0.63 (Skeem et al. (2003, p.1451), this means that there is weaker moral understanding in Black neighborhoods.

7. HONORING FINANCIAL OBLIGATIONS

"Persistent failure to honor financial obligations" is listed by the APA among the characteristics of psychopathic personality. A measure of this failure is available in the default rates on student loans. About half of American students at colleges and universities take out loans that they are required to repay after graduation, but not all graduates repay these loans. Racial and ethnic differences in loan default rates have been calculated from the data of the 1987 National Postsecondary Student Aid Study, consisting of 6,338 cases with complete information on loans, race and ethnicity, and a number of other characteristics. In the total sample, 19.2 percent defaulted on their repayments. The percentages of loan defaulters for the five major American racial and ethnic groups are shown in row 1 of Table 2.11, and show that the default rates among Blacks and Native Americans are the highest and are about three times greater than those of Whites. Hispanics, and Asians fall intermediate between these two groups and Whites.

A further index of the failure to honor financial obligations consists of poor credit ratings. These are made on the basis of records of the non-payment of debts, unacceptably late payments, and bankruptcy. A report of race differences in poor credit ratings was made by the firm Freddie Mac of 12,000 households in 1999, and the results are given in row 2 of Table 2.12. It will be seen that Blacks have the highest percentage of poor credit ratings and Whites have the lowest, while Hispanics are intermediate.

Table 2.12. Honoring financial obligations (percentages)

	DEFAULT RATES	ASIAN	BLACK	HISPANIC	NATIVE AMERICAN	WHITE	REFERENCE
1	Student loan default	34.5	55.5	20.2	44.7	15.0	Volkwein et al., 1998
2	Poor credit ratings	-	48	34	-	27	Holmes, 1999

8. DISHONESTY AND CHEATING

Dishonesty and cheating are other expressions of the moral weakness that is a central feature of psychopathic personality. Racial differences in dishonesty have been reported by Fetters, Stowe and Owings (1984) in a study in which more Black than White high school students exaggerated their self-reported grade point averages when these were checked against the records.

Racial differences in cheating in sport have been reported by Dutton and Lynn (2014) in a study that examined the percentages of Blacks and Whites who cheated in the American National Football League (NFL). They report that in the years 2000-2011, 67 percent of NFL players were Black and 31 percent were White, while of those suspended for the use of drugs to enhance performance 81 percent were Black and 18 percent were White. This study also examined cheating in the American National Basketball Association (NBA) in 2013. Seventy-eight percent of NBA basketball players were

Black, 17 percent were White, and the rest were mixed, Hispanic, or of unidentifiable race. Of those fined or suspended for cheating - including mocking, punching, flopping (an intentional fall in order to call a non-existent foul), fighting, steroid use and head-butting - 88 percent were Black. Thus, in both the American National Football League and the American National Basketball Association, Blacks were over-represented among cheaters.

9. AGGRESSIVE AND VIOLENT CRIME

Psychopathic personality is frequently expressed in crime or, as the APA expresses it, "failure to conform to social norms with respect to lawful behavior". There is an association between psychopathic personality and crime because the high aggression of those with psychopathic personality leads them to commit crimes of violence and because their weak moral sense leads them to commit non-violent crime. Numerous studies have shown that psychopathy predicts crime and recidivism (e.g. Leistico, Salekin, DeCoster & Rogers, 2008; Salekin, Rogers & Sewell, 1996). Lykken (1995) considered that more than half of those in prison are psychopaths. A review of the research on the relationship between psychopathic personality and crime by Moran (1999) concluded that around 60 percent of males in prison are psychopaths, and this figure was confirmed by Daderman and Kristiansson (2003) for 17 year olds in Sweden. A report issued by the Correctional Service of Canada (1990) estimated that approximately 75 percent of male prisoners are psychopaths. Guze (1976) estimated psychopathic personality at 79 percent for male prisoners and 68 percent for female, while Hare (1983) estimated the percentage at between 40 and 50 percent. It is difficult to assess the precise proportion of prisoners who are psychopaths because it is frequently in their interests to conceal the extent of their psychopathic personality in order to secure early release, parole, and privileges. Nevertheless it is indisputable that psychopathic personality is relatively high among criminals, and particularly among those convicted of sufficiently serious crimes to be imprisoned. We would therefore expect that racial and ethnic differences in psychopathic personality should appear in rates of crime and imprisonment.

It has been known for many decades that in the United States crime rates are high in Blacks, intermediate in Whites and low in Northeast Asians. The high crime of Blacks was shown for the 19th century in Philadelphia where the homicide rate of Blacks was three times greater than that of Whites, and in the middle of the 20th century about twelve times greater than that of Whites (Lane, 1979). The crime rate of lower status Black boys in Philadelphia in the middle of the 20th century was over three times greater than that of lower status White boys (Wolfgang, Figlio & Sellin, 1972). It was shown in the 1930s by Reid (1939) that Blacks were three to seven times over-represented in New York prisons, compared with Whites.

Crime can be either aggressive and violent, or non-violent. Race differences in aggressive and violent crime are summarized in Table 2.13. Rows 1 and 2 give rates of convictions for homicide per 10,000 population for 1979-81, and show the rate for Black men 6.4 times greater than that for White; and the rate for Black women 4.3 times greater than that for White. Row 3 gives more recent homicide rates for California for 2000-09 for ages 15 to 74, and shows the highest rate for Blacks followed by Hispanics, Native American, and Whites, and the lowest rate in Asians. Row 4 give rates of convictions for robbery per 10,000 population for 1957-88, and shows the rate for Blacks 12 times greater than that for Whites. Row 5 gives rates of husband assault on wives for 1989, and shows the rate for Blacks 3.8 times greater than that for Whites. Row 6 gives results of females in the same study and shows the rate for Blacks slightly higher at 1.21 times greater than that for Whites. The rate for Native Americans is the again highest at 1.54 times the rate for Whites. Row 7 gives results of a 1989 study in which 15-18 year olds answered a questionnaire on whether they had ever committed rape or assault, and shows the rate for Blacks more than double that for Whites.

Table 2.13. Race differences in convictions for aggressive and violent crime per 10,000

	CRIME	SEX	ASIAN	BLACK	HISP.	NATIVE AMERICAN.	WHITE	REFERENCE
1	Homicide	M	-	6.4	-	-	1.0	Lester,1989
2	Homicide	F	-	1.3	-	-	0.3	Lester,1989
3	Homicide	M/F	.20	2.09	1.32	1.23	.26	Feldmeyer. et al., 2013
4	Robbery	M/F	-	27.7	-	-	2.3	Lafree et al., 1992
5	Rape/ assault	M	-	51.6	-	68.1	48.9	Gruber et al., 1996
6	Rape/ assault	F	-	34.5	-	44.0	28.5	Gruber et al., 1996
7	Rape	M	-	9.1	-	-	3.8	Valois et al., 1999

Race differences in aggressive and violent crime, expressed as odds ratios with the rate for Whites set at 1.0, are summarized in Table 2.14. All these results show that Blacks have much the highest crime rates, while Whites have lower rates and Asians have the lowest. Hispanics and Native American rates are intermediate. A review of studies of race differences in violent recidivism by Piquero et al. (2015) has concluded that this is more prevalent among Blacks and Hispanics than among Whites.

Table 2.14. Race differences in convictions for aggressive and violent crime per 10,000 (odds ratios)

CRIME	YEAR	SEX	ASIAN	BLACK	HISPANIC	NATIVE AMERICAN	WHITE	REFERENCE
Assault	1994	M/F	0.5	5.0	3.0	2.0	1.0	Taylor & Whitney, 1999
Rape	1994	M	0.5	5.5	3.0	2.1	1.0	Taylor & Whitney, 1999
Robbery	1994	M/F	0.8	11.2	3.0	1.7	1.0	Taylor & Whitney, 1999
Murder	2001	M/F	0.2	8.2	2.4	2.3	1.0	Taylor, 2005
Manslaughter	2001	M/F	0.2	5.8	2.3	2.3	1.0	Taylor, 2005
Rape	2001	M	0.1	2.4	1.2	1.8	1.0	Taylor, 2005
Robbery	2001	M/F	0.2	14.4	4.0	3.2	1.0	Taylor, 2005
Assault	2001	M/F	0.2	7.2	3.8	3.7	1.0	Taylor, 2005

It has sometimes been asserted that race differences in convictions for crime are attributable to judges and juries being biased against Blacks. This contention has been refuted in a review of the evidence by Tonry (1994, p. 108) who writes "the conclusion that involvement in crime, not racial bias, explains much of the black disproportion among prisoners is consistent with most recent reviews of research on discrimination in sentencing." These race differences cannot be explained by the prejudice of teachers, psychiatrists, juries, judges, etc. because they are present in self-assessed questionnaires such as the psychopathy scale of the MMPI (Minnesota Multiphasic Personality Inventory) in which Black and Native Americans rate themselves as more psychopathic than Hispanics and Whites; and they are also found in victim surveys that show that victims report more attacks by Blacks than by Whites (Taylor & Whitney, 1999).

10. ALL CRIME

Race differences in imprisonment rates for all crimes, most of which do not entail violence, are summarized in Table 2.15.

Table 2.15. Race differences in imprisonment for crime per 10,000 population

AGE	SEX	ASIAN	BLACK	HISP.	NAT. AMER.	WHITE	REFERENCE
All	M/F	9.1	157.1	68.8	51.9	19.3	Taylor & Whitney, 1999
18-29	M	-	2,800	650	-	550	Unz, 2010
All	M/F	-	152	69	-	26	Harden, 2014

Row 1 gives rates of imprisonment for 1994 for the five major American racial and ethnic groups, calculated from the annual crime statistics published by the American Department of Justice. It will be seen that Blacks have the greatest incarceration rate, followed in descending order by Hispanics, Native Americans, Whites, and Asians. Row 2 presents rates of imprisonment for 2005 for men for 18-29 year olds calculated by the Bureau of Justice Statistics, which shows that Blacks have the greatest incarceration rate, followed by Hispanics. Whites have the lowest. The figures in row 2 are much higher than those in row 1 because they are for males only and for the age that has the highest crime rate. Row 3 presents rates of imprisonment for 2012 showing similar differences to those in row 1.

Race differences in convictions for non-violent crime per 10,000 presented as odds ratios are given in Table 2.16.

Table 2.16. Race and ethnic differences in convictions for non-violent crime per 10,000 (odds ratios)

	CRIME	YEAR	SEX	BLACK	ASIAN	HISPANIC	NATIVE AMERICAN	WHITE	REFERENCE
1	Car theft	1994	M/F	5.6	0.8	3.0	2.6	1.0	Taylor & Whitney, 1999
2	Burglary	2001	M/F	5.0	0.1	3.8	3.8	1.0	Taylor, 2005
3	Larceny	2001	M/F	6.3	0.2	1.9	3.2	1.0	Taylor, 2005
4	Car theft	2001	M/F	5.2	0.2	3.1	6.3	1.0	Taylor, 2005
5	Fraud	2001	M/F	4.0	0.2	1.0	0.8	1.0	Taylor, 2005
6	Drug offences	2001	M/F	12.5	0.2	5.2	1.8	1.0	Taylor, 2005
7	Drug offences	2010	M/F	2.8	-	1.6	-	1.0	Diamond et al., 2012

These studies show crime rates highest for Blacks, followed in descending order by Hispanics and Native Americans, and then by Whites and lowest in Asians. Row 7 gives rates of drug offences for

Blacks, Hispanics and Whites in an American correctional facility and show rates highest for Blacks, intermediate for Hispanics and lowest for Whites.

11. LONG-TERM MONOGAMOUS RELATIONSHIPS

A prominent feature of psychopathic personality is an inability to form long-term monogamous relationships in marriage or stable unions, listed by the American Psychiatric Association as "failure to sustain a monogamous relationship for more than one year". This failure is principally due to a reduced capacity to experience love and the need to form long term and committed relationships. Thus Lykken (1995, p. 26) writes of the psychopath's "undeveloped ability to love or affiliate with others," and Hare (1994, pp. 52, 63) writes that, "psychopaths view people as little more than objects to be used for their own gratification," and "equate love with sexual arousal."

Marriage is the most explicit expression of the willingness to enter into a committed long-term relationship based on love. Research by Forste and Tanfer (1996) has shown that couples who marry are more committed to each other than those who cohabit. As two American psychologists have written: "In its purest or ideal form, husband-wife marriage is monogamous, eternal and forsakes all other relationships." (Staples & Johnson, 1993, p.139). Several American studies have shown that Blacks attach less value to marriage than Whites. For instance, Trent and South (1992) found that Blacks are less likely than Whites to agree that "marriage is for life." Staples and Johnson (1993, p. 164) write that "Blacks do not rank marriage as highly as whites," and that "Black Americans' acceptance of this form of relationship is inconsistent with their African heritage." More recently, Stanik, McHale & Crouter (2013) have written that "African Americans experience high rates of marital discord and dissolution."

Studies of racial differences in attitudes to love and marriage are summarized in Table 2.17. Rows 1 to 4 present the results of a questionnaire study of Black and White students' attitudes to love, marriage and sex, carried out on 1,132 students at Rutgers University. Rows 1 and 2 give results for the question "I think about marriage: very frequently, frequently, sometimes, never," and show that many more Whites than Blacks gave the answers "very frequently or frequently." The results suggest that Whites are more concerned with finding a partner with whom to form a long-term relationship based on love and marriage. The second question was "I think that love and sexual intercourse should be almost always related or very related." The results given in rows 3 and 4 show a much greater endorsement of this statement by White males than by Black, suggesting a greater valuation of love by Whites, and a slightly greater endorsement by White females as compared with Black.

The results of a further study of racial differences in attitudes to marriage are shown in rows 5-8 derived from an analysis of the American National Survey of Families and Households, a national probability sample of 13,017 adults interviewed in 1987-1988. The data consist of those in the sample aged 19–35 who were unmarried and not cohabiting and had responded to the question "I would like to get married someday." The responses were scored on a strongly agree – strongly disagree scale. The results for Black, White, and Hispanic, males and females are expressed as d scores in relation to zero for Whites. The results show that Black males and females both have less desire to be married than Whites. The difference is greater for males, consistent with the results in the first two rows of the table. The results for the Hispanic males and females are inconsistent, with Hispanic males having a stronger desire to be married than Whites, but Hispanic females having a less strong desire to be married. The author of the study attributes the strong desire of Hispanic males to be married to the large numbers of them relative to Hispanic females, arising from greater numbers of male immigrants. The ratio of unmarried men to women in this age group of Hispanics was 1.38:1. This tends to make Hispanic males keen to marry in order to secure a mate. This study was also analyzed for the percentages of Blacks, Hispanics and Whites, who expressed no desire for marriage. These are given for males in row 7 and for females in row 8, showing that greater percentages of both Black males and Black females have no desire for marriage. Row 9 gives results for the Conjugal Love Scale, a questionnaire for the measurement of the strength of commitment to married love. Results for 325 Whites and for 106 Blacks show a Black d score of -0.34, indicating that Blacks have a lower commitment to married love than Whites.

Table 2.17. Attitudes to marriage

MEASURE	SEX	STAT.	BLACK	HISP.	WHITE	REFERENCE
I think about marriage frequently	M	%	3.3	-	20.2	Houston, 1981
I think about marriage frequently	F	%	29.5	-	39.7	Houston, 1981
Love and sex should be related	M	%	29.2	-	61.7	Houston, 1981
Love and sex should be related	F	%	76.3	-	79.1	Houston, 1981
I would like to get married	M	D	-0.31	0.34	0.0	South, 1993
I would like to get married	F	D	-0.16	-0.22	0.0	South, 1993
No desire for marriage	M	D	23.4	8.7	15.4	South, 1993
No desire for marriage	F	D	21.8	25.3	17.1	South, 1993
Conjugal love scale	M/F	D	-0.34	-	0.0	Munro & Adams, 1978; Philbrick et al., 1988

In addition to these studies, Broman (1993) has analyzed data of a representative sample of 2,059 married Americans and found that Blacks are less happy in marriage than Whites: "Blacks are significantly less likely to feel that their marriages are harmonious and are significantly less likely to be satisfied with their marriages" (p. 726). The propensity to form monogamous relationships based on love can also be measured by the extent to which people enter into marriage or stable cohabitation. Racial differences for these are shown in Table 2.18.

Table 2.18. Race differences married or co-habiting (percentages)

	ASIAN	BLACK	HISP.	NATIVE AMER.	WHITE	REFERENCE
1	-	47	-	39	78	Berman & Leaske, 1994; Raley, 1996
2	-	59	-	50	87	Berman & Leaske, 1994; Raley, 1996
3	66	35	55	48	63	Shi, 1999
4	79	39	65	53	68	Peng & White, 1994
5	-	36	56	-	71	Estrada-Martínez et al., 2013

Rows 1 and 2 give the percentages of Black and white men and women aged 27 and 33 who had ever been married from the National Survey of Families and Households of 1988 of 3,101 19 -34 year olds, and show much lower marriage rates among Blacks than among Whites. Shown also are the percentages married of the same age groups among Native Americans in Alaska provided by Berman and Leask (1994), showing lower rates of marriage than for Blacks and Whites. Row 3 confirms these differences from an analysis of the Medical Expenditure Panel Survey of 1997-1998 of a representative sample of 14,811 18 year olds. This survey provides data on marriage rates of Whites, Blacks, Hispanics and Asians, and marriage rates for Native Americans provided by Berman and Leask (1994) are also given. It will be seen that Asians have the highest marriage rate, followed in descending order by Whites, Hispanics, Native Americans, and Blacks. Row 4 presents data from the 1988 National Educational Longitudinal Study of approximately 25,000 14 year olds, and shows that the percentages living with both parents showed the same racial rank order, with the highest percentage being among Asians, followed again in descending order by Whites, Hispanics, Native Americans and Blacks. Row 5 gives results for 16 year olds living with both parents reported in 2008 and shows that the percentages living with both parents was highest among Whites, intermediate among Hispanics, and lowest among Blacks.

While these data are consistent with Blacks having a lower propensity than Whites to form monogamous loving relationships, an alternative explanation could be that Blacks form these relationships to the same extent as Whites but they cohabit rather than entering into formal marriage contracts. This possibility is examined in Table 2.19, and gives 1989 data for women who have either married or co-habited by ages 27 and 33. The data show that these are lower among Blacks than among Whites.

Table 2.19. Race differences in percentages of women married or co-habiting in 1989

	AGE	BLACK	WHITE	REFERENCE
1	27	70	84	Raley, 1996
2	33	80	94	Raley, 1996

A second explanation for the low marriage rates among Blacks, proposed by Wilson (1987), is that there is a shortage of young Black males making attractive marriage partners because of the large numbers of them who are either long-term unemployed or in prison. This theory has been examined by Lichter, McLaughlin,

Kephart and Landry (1992), Raley (1996) and South and Lloyd (1992), all of whom have concluded that it could explain only about one fifth of the low marriage rate among Blacks. The remaining four fifths still require explanation, and are attributable to a lesser desire to be married or cohabit among Blacks. An extreme form of the intolerance of stable monogamous unions is present when people kill their spouses. This is more common among Blacks than among Whites. Thus, in Detroit in 1982-1983 63 percent of the population was Black while 90.5 percent of those who killed their spouses were Black (Goetting, 1989).

12. EXTRAMARITAL SEX

Sexual promiscuity is one of the classical defining features of psychopathic personality. This has been confirmed in a study by Kastner and Sellbom (2012) who administered the Global Psychopathy Scale to 193 men and 200 women students, and reported significant correlations with 14 measures of sexual promiscuity including number of sexual partners in the past year (r= .42), number of one time sexual partners in the past year (r= .35), enjoying casual sex with different sexual partners in the past year (r= .46), and no need to be attached to sexual partner to enjoy sex (r=.40).

Race differences in sexual promiscuity are expressed in extramarital sex, and multiple sexual partners and are considered in this and the next section. Race differences in extramarital sex are given in Table 2.20. These show that when they do marry, Blacks are less tolerant than Whites of the monogamous constraints imposed by the marriage contract.Row 1 shows the results of an analysis of the Kinsey data of college graduates for 1938/63, in which 51 percent of Blacks were unfaithful to their spouses during the first two years of marriage compared with 23 percent of Whites. Rows 2 and 3 are for 1994 for 18-75 year olds, and show the results of a study of a representative sample of 2,172 individuals obtained in the American General Household Survey showing that for both males and females the incidence of marital infidelity was about 50 percent greater among Blacks than among Whites. Rows 4 and 5 show the results for 1990 of an analysis of the National AIDS Behavioral Survey of 1,686 married individuals who were asked if they had had extramarital sex in the last 12 months. The results show substantially greater extramarital sex among Blacks and Hispanics than among Whites. Rows 6 and 7 present the percentages of Blacks and Whites who had had extramarital sex in the last year and the last five years, found in an American national sample of 2,058 18-70 year olds, and show rates of extramarital sex three to four times more prevalent among Blacks than among Whites. In further confirmation of this difference, Moore & Schwebel (1993) have found that Blacks cite infidelity more frequently than Whites as a cause of divorce.

Table 2.20. Race differences in extramarital sex (percentages)

MEASURE	SEX	BLACK	HISPANIC	WHITE	REFERENCE
First 2 years	M/F	51	-	23	Rushton, 2000
Ever	M	33	-	21	Wiederman, 1997
Ever	F	16	-	11	Wiederman, 1997
Last year	M	4	11	2	Choi et al., 1994
Last year	F	4	5	1	Choi et al., 1994
Last year	M//F	13	-	3	Leigh et al., 1993
Last 5 years	M//F	17	-	5	Leigh et al., 1993

13. MULTIPLE SEXUAL PARTNERS

Having multiple sexual partners is a prominent of characteristic of psychopathic personality. Thus, Visser et al. (2010, p. 833) write, "promiscuity is generally considered a defining feature of psychopathy," and report correlations of .42 and .28 between psychopathy and number of lifetime sexual partners in a Canadian student sample. Studies of the racial and ethnic differences in the numbers multiple sexual partners are summarized in Table 2.2.

Table 2.21. Race differences in multiple sexual partners (percentages)

N PARTNERS	AGE	YEAR	SEX	ASIAN	BLACK	HISP.	WHITE	REFERENCE
1	6 +	-	1938/63	M/F	-	45	-	27	Rushton, 2000
2	5+	18+	1988/90	M	-	11.1	-	3.8	Seidman & Aral., 1992
3	5+	18+	1988/90	F	-	1.1	-	0.5	Seidman & Aral., 1992
4	4 +	16	1990	M	-	52	-	29	Richter et al., 1993
5	4 +	16	1990	F	-	23	-	15	Richter et al., 1993
6	4 +	15/18	1995	M	-	52	24	15	Warren et al., 1998
7	4 +	15/18	1995	F	-	22	12	13	Warren et al., 1998

N PARTNERS	AGE	YEAR	SEX	ASIAN	BLACK	HISP.	WHITE	REFERENCE
8	2 +	15/19	1995	M	-	63	49	39	Moore et al., 1998
9	2 +	15/19	1995	F	-	33	26	26	Moore et al., 1998
10	5 +	15/18	1992	M//F	8	38	21	26	Schuster et al., 1998
11	2 +	35/74	1990	M	-	-	21	22	Sabogal et al., 1995
12	2 +	35/74	1990	F	-	-	4	8	Sabogal et al., 1995
13	2 +	18/44	1988/96	M	-	40	-	22	Finer et al., 1999
14	2 +	18/44	1988/96	F	-	21	10	10	Finer et al., 1999
15	4 +	18/33	1988/90	M	-	18	-	7	Seidman & Aral, 1992
16	4 +	18/33	1988/90	F	-	7	-	1	Seidman & Aral, 1992
17	5 +	18/70	1990	M	-	16	-	9	Leigh et al., 1993
18	5 +	15/18	1990	F	-	3	-	1	Leigh et al., 1993
19	4+	15/18	1990	M	-	13	-	5	Valois et al., 1995

	N PARTNERS	AGE	YEAR	SEX	ASIAN	BLACK	HISP.	WHITE	REFERENCE
20	4+	15/18	1990	F	-	2	-	1	Valois et al., 1995
21	2+	18/75	1990	M/F	-	15	14	6	Dolcini et al., 1993
22	6 +	18/25	1991	M/F	-	7	1	3	Binson et al., 1993
23	2+	18/45	1996	F	-	30	13	14	Quadagno et al., 1998
24	4 +	15/18	1993	M	-	57	-	25	Valois et al., 1999
25	4 +	15/18	1993	F	-	26	-	15	Valois et al., 1999
27	2 +	15/19	1988	M/F	-	11	6	5	Sonenstein et al., 1991
28	3 +	12/14	2009	MF	-	14	27	5	Moore et al., 2013

Row 1 gives data abstracted by Rushton (2000) from the Kinsey archive, and shows that about twice as many Black college graduates had had six or more partners before marriage than Whites. Rows 2 and 3 give General Social Survey data showing more than twice the percentages of Black males and females than of Whites aged 18 years and older had had five or more sexual partners in the last year. Rows 4 and 5 show substantially greater percentages of Black male and female 18 year olds than of White who have had four or more sexual partners found in the American nationally representative Youth Risk

Behavior Survey of 1990. Rows 6 and 7 give the percentages of male and female Black, Hispanic and White, 15-18 year olds who had had four or more sexual partners found in the American nationally representative Youth Risk Behavior Survey of 1995. The figures show substantially greater proportions of Black males having had four or more partners than Whites and somewhat greater proportions of Black females, with Hispanics being broadly similar to Whites. Rows 8 and 9 show similar data from the American National Survey of Family Growth and National Survey of Adolescent Males. The data consist of the proportions of 15-19 year olds who have had two or more sexual partners, and show substantially greater proportions of Blacks. Row 10 gives data collected in a study of 2,026 15-18 year olds in Los Angeles for those who had had five or more sexual partners, and shows the highest percentage among Blacks and the lowest percentage among East Asians.

Rows 11 and 12 show Hispanics and Whites had approximately the same percentages of two plus partners during the last year. Rows 13 and 14 present data from the National Survey of Family Growth of approximately 20,000 women aged 18-44, and the General Social Survey of approximately 2,000 men aged 15-44. The data consist of the percentages of those who had had two or more sexual partners during the preceding year, and show that there were approximately twice as many Blacks in this category as Whites. Rows 15 and 16 give data obtained from a nationally representative sample of 4,390 for the percentages of those who had had four or more sexual partners during the last four years, and show about twice as many Blacks falling into this category as Whites. Rows 17 and 18 show the percentages of those who had had five or more sexual partners in the last year and in the last five years, obtained in an American sample of 2,058. Rows 19 and 20 show the percentages of those who had had four or more sexual partners in the last three months and show higher percentages in Blacks than in Whites. Row 21 gives results from a 1990 survey of 10,630 people aged 18 to 75, and shows that Blacks had the highest number of two plus sexual partners during the last year, followed by Hispanics. Whites had the lowest number. Row 22 presents results from the 1990 National AIDS Survey of young adults, and shows that Blacks had the highest number of six plus sexual partners, followed by Whites, and Hispanics the lowest number. Row 23 gives results for the percentages of those who had had two or more sexual partners in the last six months, and shows this was highest in Blacks and lowest among Hispanics and Whites.

Rows 24 and 25 show data from a study of 3,805 15-18 year olds and show that the percentage of Black males who had had four or more sexual partners was more than double that of Whites. Row 26 presents results from the 1988 Survey of Adolescent Males, a nationally representative study of 1,251 youths aged 15-19, and shows that Blacks had more than twice as many sexual partners than Whites and that Hispanics had slightly more sexual partners than Whites. Row 27 presents more recent results from a 2009 nationally representative study of 12-14 year olds, and shows that Blacks had more than twice as many sexual partners than Whites, while Hispanics had the highest numbers of sexual partners.

Race differences in multiple sexual partners expressed as odds ratios with Whites set at 1.0 are given in Table 2.22. This shows results from an analysis of the 1992 National Health Interview Survey of 5,223 14-22 year olds. The data are presented as odds ratios for males and females of having had sex with two or more partners during the last three months, and show that among both males and females Blacks had approximately twice as many in this category as Whites, with Hispanics falling intermediate.

Table 2.22. Race differences in multiple sexual partners (odds ratios)

MEASURE	SEX	BLACK	HISP.	WHITE	REFERENCE
2 + last 3 months	M	2.8	1.4	1.0	Santinelli et al., 1999
2 + last 3 months	F	1.4	1.1	1.0	Santinelli et al., 1999

Further studies reporting numbers of sexual partners of Blacks and Whites are given in Table 2.23. Rows 1 and 2 give data for adults for numbers of sexual partners during the last five years recorded in the surveys of the National Opinion Research Center for the years 1990-1996, and show that for both men and women Blacks had significantly more sexual partners than Whites. Row 3 gives results for numbers of sexual partners of 16 year olds recorded in Wave 1 of the National Longitudinal Study of Adolescent Health, and shows Blacks had significantly more sexual partners than Whites, with mixed race adolescents intermediate.

Table 2.23. Race differences in number of sexual partners (sample sizes in parentheses)

	SEX	BLACKS	MIXED	WHITES	REFERENCE
1	Men	2.61 (291)	-	1.89 (2644)	Lynn, 2000
2	Women	1.63 (572)	-	1.36 (3381)	Lynn, 2000
3	MF	1.8 (4271)	1.5 (116)	1.1 (10,315	Rowe, 2002

14. INTIMATE PARTNER VIOLENCE

Another measure that captures both aggression and intolerance of a stable marriage relationship is intimate partner violence, assessed by assaults by husbands on wives and by wives on husbands. A study showing that male perpetrators of physical intimate partner violence are high on antisocial personality traits has been published by Sijtsema, Baan and Bogaerts (2014). Race differences for these assaults during the last year are given in Table 2.24, together with five year rates in rows 12 and 13.

Table 2.24. Race differences in last year intimate partner violence (percentages, odds ratios)

	IPV PERPETRATOR	BLACK	HISPANIC	KOREAN	NATIVE AMERICAN	WHITE	REFERENCE
1	Husband	11.0	-	-	-	3.0	Cazenaze & Strauss, 1990
2	Wife	8.0	-	-	-	4.0	Cazenaze & Strauss, 1990
3	Husband	-	7.3	-	-	3.0	Strauss & Smith, 1990
4	Wife	-	7.8	-	-	4.0	Strauss & Smith, 1990
5	Husband	6.4	-	-	7.2	2.8	Bachman, 1992
6	Husband	8.0	-	-	12.1	4.8	Gazmararian et al., 1995
7	Wife	10.8	-	-	-	3.9	Hampton et al., 1989
8	Spouse	2.8	-	-	-	0.9	Newby et al.., 2000
9	Partner: OR	1.5	-	-	1.3	1.0	Kyriacou et al.,1999
10	Male	18.1	-	-	-	8.6	Smith et al., 2002
11	Husband	-	-	2.0	-	-	Liles et al., 2012
12	Husband	30	21	-	-	16	Caetano et al. 2000
13	Wife	23	17	-	-	12	Caetano et al. 2000

Rows 1 and 2 give rates of husband assault on wives and wife assault on husbands in 1975, and shows the rates for Blacks are two to three times greater than those for Whites. Rows 3 and 4 give rates of husband severe assault on wives and wife severe assault on husbands in 1985, and shows the rates for Hispanics about twice that for Whites. Row 5 gives a 1988 study of rates of husband assault on wives and shows the rate for Blacks 2.3 times greater than that for Whites, while the rate for Native Americans is the highest at 2.6 times the rate for Whites. Row 6 gives a 1991-2 study of rates of husband assault on wives and shows the rate for Blacks 1.7 times greater than that for Whites, while the rate for Native Americans is again the highest at 2.5 times the rate for Whites.

Row 7 gives rates of wives' assault on husbands in a 1988 study and shows the rate for Blacks 1.9 times greater than that for Whites. Row 8 gives results of a 1989 study of rates of spouse in the US military, and the rate for Blacks 3.1 times greater than that for Whites. Row 9 gives results of a 1997-8 study of rates of male partners' assault on women, with the rate for Whites set at 1.0 and shows a higher rate for Native Americans and the highest rate for Blacks. Row 10 gives results of a 2000 study of rates of male partners' assault on women, showing a rate approximately two and a half times greater for Blacks than for Whites. Row 11 gives results of a 2006 study of a representative sample of 485 Korean women aged 18-82, of whom two percent reported that they had been physically assaulted by their husbands during the last year. This figure is lower than that of the others given in the table. Rows 12 and 13 gives results of a 1995 study of intimate partner violence during the last five years for a representative national sample, showing rates for Blacks about double those for Whites, and the rate for Hispanics intermediate. In addition to these studies, lifetime and past two year prevalence rates for Black women in Baltimore of 37 and 26 percent are given by Stockman et al. (2014).

The most extreme form of intimate partner violence is homicide. Race differences in this have been given for Blacks and Whites for husbands and wives, and for boyfriends and girlfriends for 1976 and 1995 by Puzone, Saltzman, Kesno, Thomson and Mercy (2000). Their results are given in Table 2.25. It will be seen that for all eight comparisons, Blacks kill their intimate partners much more frequently than do Whites.

Table 2.25. Race differences in intimate partner homicide per 100,000

PERPETRATOR	YEAR	BLACK	WHITE
Husband	1976	11.03	0.69
Wife	1976	7.15	1.15
Husband	1995	1.98	0.25
Wife	1995	2.55	0.89
Boyfriend	1976	5.21	0.58
Girlfriend	1976	9.21	0.32
Boyfriend	1995	2.38	0.75
Girlfriend	1995	2.29	0,17

15. DELAY OF GRATIFICATION

It has been shown that an inability or unwillingness to delay immediate gratification for a long-term advantage is a component of psychopathic personality in studies by Blanchard, Bassett and Koshland (1977) and Newman, Kosson & Patterson (1999). A race difference in the delay of gratification has been noted by Banfield (1974, p. 54), who wrote of the "extreme present-orientation" of Blacks. The first study to demonstrate differences between Blacks and Whites in the delay of gratification in the United States was carried out on Black and White 9 year olds in New York City by Seagull (1966). He offered Black and White children the choice between being given a small candy bar now, or a larger one in a week's time. He found that Black children were much more likely to ask for the small candy now. This difference has been confirmed in two subsequent studies.

The first was on 15 year olds in Atlanta by Zytkoskee, Strickland and Watson (1971), and the second by Price-Williams & Ramirez (1974) was carried out on 10 year olds in the southern United States and included Mexicans as well as Blacks and Whites. The choices were varied slightly and consisted of the option of $10 dollars now or $30 in a month's time, a 5 cents candy bar now or a 25 cents bar in a month's time, and a small present now or a larger one in a month's time. There was little difference between the Black and Mexican children, both of whom preferred the immediate offer to the more distant one, as compared with Whites. A study of Black-White differences in delay of gratification among adults has been reported by Warner and Pleeter (2001), in which a total of 66,483 men in the military opted to retire in 1992 and were given the choice of receiving a lump sum or an annuity. The annual income they would receive from the annuity was considerably greater over the long term than the interest rate obtainable from the lump sum. Hence those who could delay gratification chose the annuity, and significantly more Whites than Blacks opted for the annuity.

The preference for a delay of immediate gratification for a long-term advantage is described in economics as having "a low time preference," as contrasted with "a high time preference" to denote preferring cash now to a greater sum including accrued interest in the future. These time preferences are expressed in the extent to which people save for retirement, and for future adverse contingencies such as becoming unemployed. Race differences in saving for retirement have been investigated in a 2011 survey of a representative sample of American adults aged 25-69 in full time employment and earning in excess of $40,000 a year (ING US, 2012). The results are shown in Table 2.26. Row 1 gives the sums saved in Deferred Contribution accounts accumulated for retirement and shows that these are largest in Asians (n=350), followed by Whites (n=2,750), and lowest in Hispanics (n=250) and Blacks (n=500). Row 2 gives the sums saved in other accounts, and shows that these are largest in Asians followed by Whites, and lowest in Hispanics and Blacks. Row 3 gives the percentages that pay off their credit card debts each month, thereby avoiding interest charges, and shows that these are largest in Asians followed by Whites, and lowest in Hispanics and Blacks.

Table 2.26. Race differences in savings

	SAVINGS	ASIANS	BLACKS	HISPANICS	WHITES
1	Deferred contribution	81,000	55,000	54,000	72,000
2	Other savings	70,000	43,000	48,000	64,000
3	Credit card payoffs	75%	26%	37%	53%

Another expression of the inability to delay of immediate gratification is impulsiveness. Lynam, Moffit and Stoutamer-Loeber (1993) have published a study of thirteen year olds showing Blacks are more impulsive than Whites, as assessed by teachers and by self-assessment.

16. WORK MOTIVATION AND COMMITMENT

We look next at the weak work motivation commitment component of psychopathic personality, or the "inability to sustain consistent work behavior" in the words of the American Psychiatric Association's Diagnostic Manual. The results of a number of studies of racial and ethnic differences in work motivation and commitment are summarized in Table 2.27. Row 1 shows racial and ethnic differences in "perseverance" assessed by teachers in a 1974 study in a sample of 1,337 fourth graders; perseverance was defined as "sticks to tasks until finished; if one effort to do a job is unsuccessful, tries again; and tries hard at assignments, doesn't give up easily." The results show low persistence scores by Blacks, high persistence scores by Asians, with Whites and Hispanics intermediate.

Table 2.27. Race differences in work commitment (ds, percentages)

	MEASURE	SEX	ASIAN	BLACK	HISPANIC	NATIVE AMERICN	WHITE	REFEENCE
1	Perseverance	M/F	0.38	-0.53	0.01	-	0.00	Spring et al.,1977
2	Work commitment	M/F	0.16	-0.27	0.01	-	0.00	Luzzo, 1994
3	College grades	M/F	0.04	-0.18	-0.16	-	0.00	Young, 1991
4	College grades	M/F	-	-0.53	-	-	0.00	Vars & Bowen, 1998
5	College grades	M	-	-0.33	-	-	0.00	Vars & Bowen, 1998
6	Unemployed	M	2.4	6.3	5.5	11.6	3.9	Sowell, 1978
7	Out labor force	M	-	16.3	-	16.8	6.2	Snipp, 1991
8	Out labor force	M	-	14.0	-	-	5.2	Jencks, 1992
9	Homework: hours	M/F	3.21	1.49	1.57	-	2.46	Mau & Lynn,1999
10	Homework: hours	MF	3.87	1.96	2.03	-	3.35	Mau & Lynn,1999

Row 2 shows results for a work commitment questionnaire administered to a sample of 357 college students, showing the highest work commitment among Asians followed by Hispanics and Whites, and the lowest work commitment among Blacks. Row 3 shows the grade point averages (GPAs) of college students controlling for Scholastic Aptitude Test (SAT) scores. The figures presented are the residuals of the prediction equation for GPAs from SAT scores. The negative residuals of Blacks and Hispanics indicate that these two groups obtain poorer GPAs than would be predicted from their SAT scores, while the positive residuals of Asians and Whites indicate that they obtain better GPAs than would be predicted from their SAT scores. Rows 4 and 5 show similar results from another data set and expressed as *d*s in relation to Whites.

Several American studies of the high rates of unemployment among Blacks in inner cities have concluded that a major factor responsible for these is an unwillingness to work. For instance, Anderson (1980, p. 75) writes that "there are many unemployed black youths who are unmotivated and uninterested in working for a living, particularly in the dead-end jobs they are likely to get." Petterson (1997, p. 605) writes in similar vein that "it is commonly contended that young black men experience more joblessness than their white counterparts because they are less willing to seek out low paying jobs." American Asians are the opposite of Blacks in this respect. They have low rates of unemployment, and it has been shown by Flynn (1991) that they achieve higher educational qualifications and earnings than would be predicted from their intelligence, suggesting that they have strong work motivation. Row 6 presents data from the 1970 American census for unemployment rates for the five major racial and ethnic groups, and shows the highest rate of unemployment among Native Americans followed in descending order by Blacks, Hispanics, Whites and East Asians, consisting of ethnic Chinese and Japanese. These differences in unemployment rates are frequently attributed to White racism but it is difficult to reconcile this explanation with the lower rate of unemployment among East Asians as compared with Whites, and also with the higher rate of unemployment among Native Americans as compared with Blacks. Row 7 shows further evidence for racial and ethnic differences in work commitment expressed in the percentages of Black, Native American, and White men aged 25-54 "out of the labor force," a category consisting of healthy individuals who have not worked at all for the preceding year and whom Jencks (1992) contends cannot be looking for work and should be regarded as choosing not to work. The percentages of Blacks and Native Americans out of the labor force are approximately the same and about two and a half times greater than that of Whites. Row 8 shows data presented by Jencks (1992) for Blacks and Whites out of the labor force for 1985-1987, again showing that about two and a half times as many Blacks fall into this category as Whites. Rows 9 and 10 give data for the hours of homework done per week by tenth and twelfth graders calculated from the American National Educational Longitudinal Study, and show that Blacks and Hispanics do less homework than Whites, while Asians do more.

17. RECKLESSNESS

We examine next some expressions of recklessness included in the APA's list of the manifestations of psychopathic personality as "recklessness regarding one's own or others' personal safety." A weaker form of recklessness is risk taking which has frequently been noted as a characteristic of psychopathic personality (e.g. Dean et al., 2013) and of criminality (Ellis, Hoskin et al., 2014). Psychopaths appear to enjoy risk taking partly because they lack foresight of the likely adverse consequences, they lack inhibitory controls, and because it provides excitement. We begin this section by giving the results from the 1989-1993 American Teenage Attitudes and Practices Survey, a study of a nationally representative sample of 9,135 youths aged 12-18 years. As part of this study the youths were asked to consider the question: "Do you get a kick out of doing things every now and then that are a little risky or dangerous?'" It was found that 56.9 percent of Blacks agreed with this statement, as compared with 38.6 percent of Whites (Flint, Yamada & Novotny, 1998). It appears that Blacks enjoy risk taking more than Whites.

A number of studies showing race differences in recklessness, assessed as the non-use of seat belts in automobiles, are summarized in Table 2.28. Row 1 gives the results of a study carried out in the early 1980s in North Carolina, in which observers recorded whether people were wearing seat belts while they were driving. More women than men were found to wear seat belts, consistent with many studies showing that psychopathic personality is less common in females than in males (e.g. Dahlstrom et al., 1986), but more striking was the greater non-use of seat belts by Blacks than by Whites. Row 2 shows the results of a study of 1,063 drivers in Harrisburg, Pennsylvania, and excluded those whose automobiles had automatic seat belt devices. The results for Blacks and Hispanics were combined into one group, the figure for which is assigned to both groups in the table and is about 25 percent greater for Blacks and Hispanics than for whites. Row 3 gives data for Nevada consisting of all motor vehicle accidents causing serious injuries to children during 1989-1992; it reports the percentages of the injured children not wearing any kind of seat belt or safety harness, and found this much higher among Blacks than among Whites and Hispanics. Row 4 gives data from the Adolescent Health Survey of 13,454 Native American 18 year olds for those who never or hardly ever us ed belts. Rows 5 and 6 give the results of a study of 4,896 people carried out in 2000 at four cities in which observers recorded whether men and women were wearing seat belts while they were driving. For both men and women, fewer Blacks wore seat belts, and Hispanics wore seat belts marginally less than Whites. More women than men were found to wear seat belts. Row 7 gives data for parents putting their young children into seat belts showing that Whites used them significantly more than Blacks. In addition to these studies, Shin, Hong and Waldron (1999) report results from a study of 15-20 year olds in an American city for seat belt usage, and state that Asians used them significantly more than Whites and that Whites used them significantly more than Blacks, but they do not provide figures.

Table 2.28. Race differences in non-use of seat belts (percentages)

	SEX	BLACK	HISP.	NAT. AMER.	WHITE	REFERENCE
1	M/F	94	-	-	72	Hunter et al., 1986
2	M/F	33	33	-	25	Colon, 1992
3	M/F	95	48	-	53	Niemeryk et al., 1997
4	M/F	-	-	44	-	Blum et al., 1992
5	M	63	48	-	47	Wells et al. 2002
6	F	41	35	-	34	Wells et al. 2002
7	M/F	35	-	-	15	Macy et al., 2014

Table 2.29 gives further measures of risk taken from studies in the 1990s for males and females combined, and its more extreme expression as recklessness. Row 1 presents data from a study of 5,112 observations of red light running in three cities in Virginia, and shows that Blacks run red lights more than Whites. Row 2 presents data from a study of 429 US army personnel who had suffered motor vehicle injuries requiring hospitalization in 1992. The authors of the study found that the main factors responsible for the injuries were heavy drinking, speeding, and non-use of seat belts. Race differences in the rate of injuries were calculated in relation to total numbers among approximately 100,000 military personnel, and showed that Blacks sustained 78 percent more injuries than Whites. Row 3 gives the results of a study of motor vehicle accidents requiring hospitalization among Native Americans in the state of Washington over the years 1990-1994 compared to all residents, and shows an excess accident rate of 82 percent. The great majority of the population of Washington state are White and are entered as such in the table.

Row 4 gives data from a study of age-adjusted fatality rates from vehicle accidents in New Mexico in the years 1958-1990, and shows that Native Americans had by far the highest rate, followed by Hispanics, and Whites having the lowest rate. Row 5 gives data from a study of age-adjusted fatality rates from vehicle accidents in Arizona in the years 1990-96, and shows that Native Americans had by far the highest rate, followed by Blacks, with Whites a little lower, and Hispanics having the lowest rate. Row 6 gives results from a study of driving after drinking excessive alcohol and shows that Hispanics had the highest rate, followed by Blacks, and Whites the lowest rate. In these studies Blacks and Native Americans showed greater recklessness and risk taking than Whites, with Hispanics generally intermediate except for their high risk-taking shown in row 6.

Table 2.29. Race differences in recklessness in automobile driving (percentages and odds ratios)

	MEASURE	BLACK	HISP.	NAT. AMER.	WHITE	REFERENCE
1	Red light runs: OR	1.19	-	-	1.0	Porter & England, 2000
2	Injuries: OR	1.78	-	-	1.0	Bell et al., 2000
3	Injuries: OR	-	-	1.82	1.0	Sullivan & Grossman, 1999
4	Fatalities: %	-	.007	.023	.005	Schiff & Becker, 1996
5	Fatalities: %	.013	.010	.052	.012	Campos-Outcalt et al., 2003
6	Drink driving: %	9	15	-	7	Voas et al., 1998

A further expression of recklessness is pathological gambling. The American Psychiatric Association's Diagnostic and Statistical

Manual of Mental Disorders lists pathological gambling as a disorder of impulse control, which is itself a central component of psychopathic personality.

Table 2.30. Race differences in gambling (percentages)

	GAMBLING	ASIAN	BLACK	HISPANIC	NATIVE AMERICAN	WHITE	REFERENCE
1	Problem	-	-	-	14.8	9.6	Zitzow, 1996a
2	Pathological	-	-	-	2.8	1.6	Zitzow, 1996b
3	Problem	-	-	-	12	6	Cozzetto & Larocque, 1996
4	Problem	-	-	-	7.1	2.5	Volberg & Abbott, 1997
5	Problem	-	-	-	9.9	3.3	Peacock et al., 1999
6	Weekly	4	22	22	30	5	Stinchfield, 2000
7	Pathological	-	-	7.9	-	1.8	Welte et al., 2001
8	Daily	-	-	9.5	-	4.0	Stinchfield, 2007
9	Pathological	-	2.2	1.0	-	1.2	Alegria et al., 2009
10	Pathological	-	.96	-	-	.45	Barry et., 2011a
11	Pathological	-	-	.4	-	.5	Barry et., 2011b
12	Never	25	-	-	-	15	Kong et al., 2013

In a study of 140 drug-abusing anti-socials, 29 percent were found to be pathological gamblers (Lesieur et al., 1986), while Argo and Black (2004) reported that between 15 to 40 percent of pathological gamblers were antisocial personalities. In a review of research, Derevensky (2008, p.411) has written that pathological gamblers "have difficulty in conforming to social norms and experience difficulties with self-control, are more impulsive, and are greater risk-takers… and have a greater frequency of attention deficit hyperactivity disorder and conduct problems," all of which are expressions of psychopathic personality. In another review of research it was shown that problematic gambling is associated with delinquency (Johansson, Grant, Kim, Odlaug & Götestam, 2009). Table 2.30 summarizes studies of race differences in gambling. The first five rows show rates of problem gambling and more severe pathological gambling substantially higher in Native Americans than in Whites. Row 6 confirms this for weekly gambling, showing the highest rate in Native Americans, followed by Blacks and Hispanics, and much lower rates in Asians and Whites. Rows 7 and 8 confirm higher rates of gambling in Hispanics than in Whites, but Hispanics have approximately the same rates of gambling as Whites in the results given in row 9. Row 10 shows pathological gambling about twice as frequent in Blacks than in Whites. Row 11 shows pathological gambling about equally frequent in Hispanics as in Whites. Row 12 shows a higher percentage of never gamblers among Asians than among Whites.

18. RECKLESSNESS IN SEXUAL BEHAVIORS

Further expressions of recklessness can be obtained from a variety of sexual behaviors. We consider first the non-use of contraception by those who do not wish to have children. This can be regarded as reckless both because this is likely to result in an unwanted pregnancy and also because it incurs the risk of contracting sexually transmitted diseases including HIV and AIDS. A number of studies of racial differences in the non-use of contraception are summarized in Table 2.31.

Table 2.31. Race differences in the non-use of contraception (percentages)

	MEASURE	SEX	ASIAN	BLACK	HISPANIC	NATIVE AMERICAN	WHITE	REFERENCE
1	First sex	M	-	66	-	-	54	Zelnik & Shah, 1983
2	First sex	F	-	59	-	-	49	Zelnik & Shah, 1983
3	First sex	F	-	37	54	-	24	Abma et al., 1998
4	First sex	F	-	21	26	-	15	Peterson et al., 1998
5	First sex	M/F	-	64	67	-	42	Schuster et al., 1998
6	First sex	F	-	66	-	-	48	Kahn et al., 1990
7	First sex	M	-	-	-	60	-	Blum et al., 1992
8	First sex	F	-	-	-	50	-	Blum et al., 1992
9	First sex	M	-	51	43	-	34	Sonenstein et al. 1989
10	First sex	M	-	64	77	-	45	Forrest & Singh, 1990
11	First sex	F	-	46	46	-	31	Forrest & Singh, 1990
12	Current: 1976	F	-	20	-	-	11	Stephen et al., 1988
13	Current: 1982	F	-	16	-	-	9	Stephen et al., 1988
14	Current: 1989	F	-	65	-	-	35	Wyatt, 1991

	MEASURE	SEX	ASIAN	BLACK	HISPANIC	NATIVE AMERICAN	WHITE	REFERENCE
15	Last year	M/F	-	27	-	-	15	Leigh et al. 1993
16	Last month	F	-	17	23	-	10	Kraft & Coverdill, 1994
17	Last sex	F	-	28	37	-	10	Darroch et al., 1999
18	Ever: age 15/17	F	-	31	-	-	21	Zelnik & Kim, 1982
19	Ever: age 18/19	F	-	29	-	-	17	Zelnik & Kim, 1982
20	Ever	F	-	43	-	-	35	Swenson e al., 1989
21	Ever	M	-	45	50	-	43	Catania et al., 1992
22	Ever	F	-	68	72	-	49	Catania et al., 1992
23	Last 6 months	M	-	52	-	-	18	Peterson et al., 1998
24	Ever	F	-	23	35	-	19	Guttmacher, 1994
25	None	M	-	48	55	-	46	Catania et al., 1994
26	None	F	-	60	76	-	48	Catania et al., 1994
27	Inconsistent	F	-	21	-	-	15	Peterson et al., 1998
28	Inconsistent	M/F	-2.64	0	-1.9	-	-.9	Sikkema et al., 2004

MEASURE	SEX	ASIAN	BLACK	HISPANIC	NATIVE AMERICAN	WHITE	REFERENCE
29 None	F	-	16	9	-	9	Mosher & Jones, 2010
30 None	F	21	45	38	-	24	Rocca, 2012

Rows 1-11 give data for the percentages of different racial groups who did not use contraception on the occasion of their first sexual intercourse. Rows 1 and 2 show the results of an analysis of an American national probability sample survey of 670 15–19 year old young women and 936 17–21 year old young men carried out in 1979. The data consist of the percentages of Blacks and Whites who did not use any type of contraception on the occasion of their first sexual intercourse, and shows lower use of contraception by Blacks than by Whites. Row 3 gives data for a national sample of 2042 15-24 year-old women who did not use any form of contraception on the occasion of their first sexual intercourse, and shows greater non-use of contraception by Blacks and Hispanics, as compared with Whites. Rows 4, 5 and 6 present data on the non-use of contraception for further samples, and rows 7 and 8 present data for a sample of 18-year-old male and female Native Americans drawn from the Adolescent Health Survey showing high rates of non-use of contraception. Row 9 shows data for 1,880 youths from the 1988 National Survey of Adolescent Males. Rows 10 and 11 give data from the 1982 and 1988 National Surveys of Family Growth based on 7,969 and 8,450 15-44 year-old women. These studies show that Blacks, Hispanics, and Native Americans are less likely than Whites to use contraception on the occasion of their first sexual intercourse. Notice, however, that Blacks, Hispanics, and Native Americans, tend to have their first sexual intercourse at younger ages than Whites, and are therefore less mature and less likely to use contraception for this reason.

Rows 12 and 13 give results for women surveyed in 1976 and 1982 described as "not pregnant, sterilized of seeking to get pregnant who, for whatever reason, have chosen not to use a contraceptive method," (p. 59) and shows that the percentages of these were greater among Blacks than among Whites. Row 14 confirms the greater percentage of Black women not using contraception. Rows 15 through 24 show the results of studies on the inconsistent or non-use of contraception on occasions other than first sexual intercourse. Row 15 shows data from an American national probability survey of 2,058 18-70 year olds, and gives the percentages of Blacks and Whites who had had more than one sexual partner in the previous year, and who never used condoms or who used them inconsistently. Row 16 gives the results from the National Longitudinal Study of Labor Force Experiences of sexually active unmarried women, showing highest rates of non-use of contraception among Hispanics, intermediate rates in Blacks, and lowest rates in Whites.

Row 17 shows the results from the 1995 National Study of Family Growth of 18 year old females who did not use contraception during their last sexual intercourse showing much higher rates in Blacks and Hispanics than in Whites. Rows 18 and 19 give data from a 1979 survey showing more Black than White females aged 15-17 and 18-19 who did not use contraception. Row 20 shows more Black than White females not using contraception. Rows 21 and 22 show data from a survey of 1,229 sexually active unmarried men and women in San Francisco with a high rate of sexually transmitted diseases who seldom or never used condoms showing higher rates in Blacks and Hispanics than in Whites. Row 23 gives results from a study of gay and bisexual African-American men who reported a substantially higher prevalence of unprotected anal intercourse during the past 6 months (52 percent) in 1990 than did gay and bisexual White men in the AIDS Behavioral Research Project (15 percent) and the San Francisco Men's Health Study (20 percent) in 1988. The authors comment that, "These results suggest that, in the second decade of the AIDS epidemic, behavioral interventions are urgently need to help African-American men reduce their high-risk behaviors."

Row 24 gives results from a study of 15-19 year old women, showing that Hispanics had the greatest percentage not using contraception followed by Blacks, and the lowest percentage in Whites. Rows 25 and 26 give results confirming that Hispanics had the greatest percentage not using contraception followed by Blacks, and Whites the lowest percentage. Row 27 shows the results from the National Survey of Family Growth for 1,485 women who took the contraceptive pill but had not taken it consistently during the preceding three months showing a higher percentage in Blacks than in Whites. Row 28 shows the results for risk taking by non-use of contraception in a sample of 1,072 teenagers with an average age of 14.5 years. The results are presented as a multinomial logistic regression analysis of higher (=1) vs no risk (=0) teens. The zero for Blacks indicates the highest score for high risk, and the negative scores for the other groups indicate lower scores for risk, with Asians obtaining the lowest score. Row 29 shows the results for risk taking by non-use of contraception in the 2006-2008 National Survey of Family Growth study of sexually active unmarried women showing that Blacks had the greatest percentage not using contraception followed by Hispanics and Whites. Row 30 gives the results for risk taking by non-use of contraception in a sample of 602 sexually active unmarried women aged 18-29, showing that Blacks had the greatest percentage not using contraception followed by Hispanics and Whites, while Asians had the lowest percentage of risk takers.

A further measure of recklessness in sexual behavior is available in failure to use contraception by those not wishing to become pregnant, resulting in unwanted and unplanned pregnancies and births. Virtually all teenage pregnancies can be regarded as resulting from recklessness in having unprotected sex because hardly any teenagers have babies intentionally (Kalmuss, 1992). Hence the prevalence of these provides a further index of recklessness in sexual behavior. Results of studies of teenage births are given in Table 2.32.

Table 2.32. Race differences in teenage females' unplanned pregnancies and births (per centages)

MEASURE		ASIAN	BLACK	HISPANIC	NATIVE AMERICAN	WHITE	REFERENCE
1	Births aged 10-14	-	0.50	0.20	-	0.05	Guttmacher, 1994
2	Births aged 10-14	-	0.30	0.26	-	0.03	Taylor et al., 1999
3	Births aged 15-23	-	8.4	6.5	-	2.3	Guttmacher, 1994
4	Births aged 15-17	-	7.7	5.2	-	2.2	Hollander, 1996
5	Births aged 15-19	2.5	8.5	7.5	-	3.5	Burke, 1997
6	Births aged 10-18	2.6	16.2	10.1	8.4	5.2	Moore et al., 1998
7	Teen pregnancy	-	27	34	31	19	Kenny et al, 1997
8	Unplanned pregnancy	-	72	49	-	43	Henshaw, 1998
9	Unplanned births	-	51	30	-	27	Henshaw, 1998

Row 1 shows data for 1990 for the percentages of girls aged 10-14 who had babies, and shows that this was ten times greater among Blacks than among Whites and four times greater among Hispanics, as compared with Whites. Row 2 presents the percentages of 10-14 year old girls who gave birth in California in 1993-1995, and shows similar results with Blacks having babies ten times more frequently and Hispanics about nine times more frequently than Whites. Row 3 shows the percentages of 15-23 year old girls having babies for 1990, and again shows that this was considerably higher for Blacks than for Whites, with Hispanics intermediate. Row 4 confirms this difference for 15-17 year olds. Row 5 gives the teenage birth rates in California in 1985, and shows the highest rate among Blacks followed in descending order by Hispanics, Whites and East Asians. Row 6 presents data for 1992 from a national sample of 18 year old women, and shows that the percentages who had had a child were more than three times greater among Blacks and about twice as great among Hispanics, about 50 percent greater among Native Americans, and about half as great among East Asians, as compared with Whites. Row 7 presents data for 1,937 young women aged 18-22 from a south-western American state, and reports the percentages who had had a teenage pregnancy, showing that this was substantially higher among Blacks, Hispanics and Native Americans, than among Whites. Rows 8 and 9 show 1994 data for the numbers of women aged 15-44 who had at least one unplanned pregnancy and at least one unplanned birth, and shows that both of these were greatest among Blacks and lowest among Whites, with Hispanics intermediate.

A study of sexual risk taking among women defined as having a partner who was HIV positive, an injection drug user in the last five years, non-monogamous, a transfusion recipient or a haemophiliac has been reported by Grinstead, Faigeles, Binson & Eversley (1993). The percentages of these sexual risk takers were Blacks 6.3, Hispanics 5.1, and Whites 4.3. Unplanned pregnancies and births are also caused by reckless behavior by men. Studies of men's contribution to unplanned pregnancies and births in the 1980s are summarized in rows 1 to 6 of Table 2.33. All the studies show greater percentages of unplanned pregnancies and births among Blacks than among Whites.

Table 2.33. Race differences in teenage males' unplanned pregnancies and births (percentages)

	MEASURE	BLACK	HISPANIC	WHITE	REFERENCE
1	Teen pregnancy	14	7	5	Marsiglio, 1987
2	Fathers pleased	9	4	4	Marsiglio, 1987
3	Fathers pleased	9	5	6	Marsiglio, 1993
4	Fathers didn't care	11.1	-	5.6	Moore & Stief, 1991
5	Fathers didn't think	16.5	-	9.2	Moore & Stief, 1991
6	Teen pregnancy	31	41	8	Thornberry et al., 1997

Row 1 shows data for teenage males obtained from a nationally representative sample of 12,686 respondents in the American National Survey of Adolescent Males. It shows that the percentage of teenage Blacks who had fathered an illegitimate child was approximately three times greater than that of Whites, with Hispanics intermediate. Row 2 gives data for adolescent males in the same survey who were asked if they would feel "very pleased" if they were responsible for an unplanned pregnancy. To be very pleased about producing an unplanned pregnancy in a teenage girl can be regarded as an index of a reckless attitude towards the well being of others. The results show that the percentage of teenage Blacks who had fathered a child and who would be "very pleased" to produce a pregnancy was more than double the percentages of Hispanics and Whites. Row 3 presents further data from a 1988 survey for adolescent males who were asked if they would feel "very pleased"' if they were responsible for an unplanned pregnancy and again shows that the percentage of teenage Blacks who had fathered a child and who would be "very pleased" to produce a pregnancy was greater than the percentages of Hispanics and Whites. Rows 4 and 5 give data for

934 youths from the 1987 National Survey of Children who were asked how they felt about their partners becoming pregnant when they first had sex with them. Row 4 gives the percentages of those who "didn't care," and shows about twice as many of these among Blacks as among Whites. Row 5 gives the percentages of those who "didn't think," and shows the percentage of these about 80 percent higher among Blacks than among Whites. Row 6 gives results for teenage males who had fathered an illegitimate child and shows this was greatest for Hispanics followed by Blacks, and lowest for Whites.

19. SEXUALLY TRANSMITTED DISEASES

It has been shown that high rates of sexual risk taking are associated with high rates of sexually transmitted diseases (STDs) (Birthrong & Latzman, 2014). Psychopaths are high sexual risk takers, so it is not surprising that it has also been shown that psychopathic personality is associated with the contraction of STDs. For instance, in a sample of 164 Black female adolescents in Baltimore, ratings for conduct disorder (a precursor of psychopathic personality) was correlated at .25 with having contracted an STD (Bachanas, Morris, Lewis-Gess et al., 2002). In a study of drug takers who were HIV infected, it was reported that 17.6 percent had antisocial personality disorder while 7.8 percent were without antisocial personality disorder (Brooner et al., 1993).

There are three reasons for the association of psychopathic personality with the contraction of STDs. First, psychopaths are risk takers so they tend to engage in unprotected and promiscuous sex, with the result that they are likely to contract STDs. Second, psychopaths have short time horizons, so they tend to opt for immediate gratification without regard to the longer term adverse consequences of possible STD infection that can be incurred by unprotected and promiscuous sex. Third, psychopaths tend to become drug addicted and are prone to engage in unprotected and promiscuous sex when they are under the influence of drugs. Thus, it has been found that women who report excessive alcohol and other drug use are significantly less likely to use condoms in their sexual relationships (Wingood & DiClimente, 1998). A correlation of .66 has been reported between drug use and having contracted an STD in a sample of Black female adolescents in Baltimore (Bachanas, Morris, Lewis-Gess et al., 2002). In addition, female psychopaths sometimes sell sex for cash to buy drugs, so they are at risk of contracting STDs. Amaro et al. (2001, pp.436-8) summarize numerous studies and conclude that "research indicates that illicit substance abuse increases women's risk for HIV," that "sexual risk-taking behaviors such as non-use of condoms, recent and past sexually transmitted diseases, multiple partnering, and sex with high risk partners, are more common among black and Latina women," and that, studies with young African-American men reveal that those who use crack are more likely to report multiple sex partnering and non-use of condoms, which are known HIV risk factors; they are more likely to engage in injection drug use as well as being more likely to engage in sex with injection drug users; they also report being more likely to engage in sex trade for drugs or money and more likely to have unprotected sex during sex trade; among incarcerated adolescents, crack users are more likely to have engaged in anal sex and frequent sex with female partners.

Race differences in the incidence and prevalence of HIV/AIDS are given for males and females combined in Table 2.34. Row 1 gives results for HIV infection per 1,000 among 137,209 Jobs Corps students aged 16-21 years in 1990 showing the highest rate for Blacks, followed by Hispanics, and the lowest rates for Whites. Row 2 gives results for AIDS given by Rushton (2000, p. 181) from data from the Centers for Disease Control and Prevention showing similar figures to those in row 1. Rows 3 and 4 give HIV incidence rates (new infections) for 2008 and 2010 and show the highest rates for Blacks, followed by Hispanics, and the lowest rates for Asians and Whites. Row 5 gives national data for the percentages of HIV incidence rates for 2011 for young men who have sex with men and shows the highest rate for Blacks, who were about 15 percent of the population and comprised 63 percent of HIV incidence rates. Hispanics and Whites were about equally represented but Hispanics were a smaller percentage of the population.

Table 2.34. Race differences in incidence and prevalence of HIV/AIDS per 1,000

	STD	ASIAN	BLACK	HISPANIC	NATIVE AMERICAN	WHITE	REFERENCE
1	HIV, 1990	-	5.3	2.6	-	1.2	St. Louis et al., 1991
2	AIDS, 1993	0.00	3.56	3.56	-	0.86	Rushton, 2000
3	HIV, 2008	0.10	0.74	0.26	0.14	0.08	Centers Disease Control, 2012
4	HIV, 2010	0.08	0.69	0.28	0.11	0.09	Centers Disease Control, 2012
5	HIV, 2011: %	-	63	16	-	18	Centers Disease Control, 2012

Race differences in sexually transmitted diseases other than HIV/AIDS are given in Table 2.35. Rows 1 through 4 show from 1982 through 2011 the highest rates of syphilis among Blacks followed by Hispanics and Native Americans, with lower rates among Whites and Asians. Rows 5 and 6 show the same race differences in 2011 for chlamydia and gonorrhea. It is possible that the race differences in STDs are determined at least in part by differences in intelligence. Low intelligence is associated with low general knowledge, so possibly Blacks and Hispanics with lower IQs have not acquired the knowledge that unprotected sex incurs the risk of STDs, as a result of which they are more prone to engage in unprotected sex and contract an STD. However, Bachanas, Morris, Lewis-Gess et al. (2002) in their study of Black female adolescents in Baltimore found no association between knowledge of HIV and having contracted an STD (r= .05). This suggests that the race differences in STDs are not determined by differences in intelligence but by differences in psychopathic personality.

Table 2.35. Race differences in incidence of sexually transmitted diseases per 100,000

Note: # = rates lower than those of whites

	STD	SEX	ASIAN	BLACK	HISPANIC	NATIVE AMERICAN	WHITE	REFERENCE
1	Syphilis, 1982	M	7.4	101.9	43.8	21.4	10.3	Rolfs & Nakashima, 1990
2	Syphilis, 1989	M	2.8	147.4	27.3	4.9	3.2	Rolfs & Nakashima, 1990
3	Syphilis, 2011	M	3.1	27.0	8.5	5.0	4.4	Centers Disease Control, 2011
4	Syphilis, 2011	F	0.1	5.1	0.6	0.5	0.3	Centers Disease Control, 2011
5	Chlamydia, 2011	MF	#	1,194	384	648	159	Centers Disease Control, 2011
6	Gonorrhea, 2011	MF	#	427.3	53.8	115.7	25.2	Centers Disease Control, 2011

20. SEXUAL PRECOCITY

It has frequently been shown that early sexual experience is a feature of psychopathic personality disorder (e.g. Khurana, Romer, Betancourt et al., 2012). Studies providing data on race differences in age of first sexual intercourse as a measure of sexual precocity are summarized in Table 2.36.

Table 2.36. Race differences in age of first sexual intercourse (percentages, ORs)

	AGE	SEX	ASIAN	BLACK	HISPANIC	NATIVE AMERICAN	WHITE	REFERENCE
1	17	M/F	-	51	-	-	21	Rushton, 2000
2	12	F	.03	.19	.08	-	.03	East, 1998
3	13	M	-	49	-	-	18	Coker et al., 1994
4	13	F	-	12	-	-	3	Coker et al., 1994
5	13	M/F	4	33	15	-	14	Schuster et al., 1998
6	13	M	-	56	52	-	-	Raine et al., 1999
7	13	F	-	21	13	-	-	Raine et al., 1999
8	15	F	-	31	-	-	11	Zelnik & Kantner, 1980
9	15	F	-	39	-	-	13	Zelnik & Kantner, 1980
10	15	F	-	41	-	-	18	Zelnik & Kantner, 1980
11	15	F	-	39	-	-	14	Hofferth et al., 1987a

	AGE	SEX	ASIAN	BLACK	HISPANIC	NATIVE AMERICAN	WHITE	REFERENCE
12	15	F	-	42	-	-	18	Hofferth et al., 1987b
13	15	M	-	65	-	-	32	Zelnik & ShaH, 1983
14	15	F	-	37	-	-	14	Zelnik & ShaH, 1983
15	15	F	-	45	-	-	14	Sonenstein et al., 1991
16	15	F	-	25	-	-	15	Hofferth & Hayes, 1987
17	15	M	-	42	19	-	12	Hofferth & Hayes, 1987
18	15	F	-	10	4	-	5	Hofferth & Hayes, 1987
19	15	M	-	69	32	-	26	Sonenstein et al., 1989
20	15	M	-	49	-	-	21	Sonenstein et al., 1991
21	15	M	-	-	59	-	32	Sabogal et al., 1995
22	15	F	-	-	15	-	12	Sabogal et al., 1995
23	15	M	-	46	38	-	29	Gruber et al., 1996
24	15	F	-	33	29	-	22	Gruber et al., 1996
25	15	M	-	-	-	22	-	Blum et al., 1992
26	15	F	-	-	-	15	-	Blum et al., 1992
27	15	MF	30	68	57	72	55	Guttmacher et al., 1997

	AGE	SEX	ASIAN	BLACK	HISPANIC	NATIVE AMERICAN	WHITE	REFERENCE
28	15/17	F	-	44	36	-	30	Forest & Singh, 1990
29	15/17	F	-	50	36	-	36	Forest & Singh, 1990
30	15/19	F	-	63	-	-	37	Ford et al., 1981
31	15/19	M	-	81	60	-	57	Sonenstein et al., 1998
32	16	M	-	57	-	-	24	Furstenberg et al., 1987
33	16	F	-	38	-	-	17	Furstenberg et al., 1987
34	16	MF	7	39	18	-	23	Durbin et al., 1993
35	16	M	24	78	68	-	51	Schuster et al., 1998
36	16	F	30	65	46	-	48	Schuster et al., 1998
37	16	M/F	27	-	-	-	50	Hou et al., 1997
38	17	M	-	42	35	-	24	Leigh et al., 1994
39	17	F	-	40	13	-	28	Leigh et al., 1994
40	17	M	-	4.6	1.2	-	1.0	Santelli et al., 2000
41	17	F	-	1.6	0.7	-	1.0	Santelli et al., 2000
42	15	M/F	0.3	2.0	1.6	-	1.0	Boyer et al., 1999
43	15/19	M	-	80	61	-	49	Sonenstein et al., 1998
44	12/14	MF	-	31	23	-	16	Moore et al., 2013

Row 1 gives Rushton's (1995) analysis of the Kinsey archive for the percentages of Black and White college students and graduates who had had sexual intercourse before the age of 17, and shows more than double the proportion among Blacks. These data were collected over the years 1938-1963. More recent studies have confirmed and extended this difference by showing greater proportions of Blacks than Whites with precocious sexual experience and providing information for other racial and ethnic groups. Row 2 gives results for a study of 527 12 year olds in California and shows that Blacks had the highest percentage who had had sexual intercourse followed by Hispanics, while Asians and Whites had the lowest. Rows 3 and 4 give data derived from a sample of 5,478 adolescents in the Youth Behavioral Risk Survey obtained in South Carolina; they give the percentages of boys and girls who had had sexual intercourse before the age of 13 and show that this was about two and a half times more frequent among Blacks than among Whites. Row 5 presents the results of a survey of 2,026 adolescents in Los Angeles who were sexually experienced by the age of 13; it confirms that the proportion of Blacks was about two and a half times greater than that of Whites; it shows also that the proportion of Hispanics was about the same as that of Whites while the proportion of East Asians was only about a third of that of Whites. Rows 6 and 7 show that Blacks had a higher percentage than Hispanics of those who had had sexual intercourse by the age of 13. Rows 8 through 27 give further studies showing the percentages of those who had had sexual intercourse by the age of 15. Rows 28 and 29 show Black females had the highest proportion of sexually experienced by the age of 15 to 17, followed by Hispanics and Whites.

Rows 30 and 31 show Blacks had the highest proportion of sexually experienced by the age of 15 to 19, followed by Hispanics and Whites. Rows 32 to 36 show Blacks had the highest proportion of sexually experienced by the age of 16, followed by Whites and Hispanics, while Asians had the lowest proportion. Row 37 gives results for the percentages of those who had had sexual intercourse by the age of 16 showing this is much lower in Asians than in Whites. Rows 38 and 39 give results for the percentages of those who had had sexual intercourse by the age of 17 showing this highest in Blacks.

Rows 40 through 42 give results as odds ratios with Whites equal to 1.0 from the Youth Risk Behavior Survey, showing higher percentages of Blacks who had had sexual intercourse by the age of 17, and about equal percentages of Hispanics and Whites. Row 43 shows Blacks had the highest proportion of sexually experienced by the age of 15 to 19, followed by Hispanics and Whites. Row 44 gives results from a 2009 study showing that among 12-14 year olds approximately twice as many Blacks as Whites were sexually experienced, with Hispanics intermediate.

The general pattern in these results is that the percentage of those with sexual experience is consistently greater among Blacks than among Whites. Hispanics are typically intermediate between Blacks and Whites while Asians are have less experience than Whites. The data for Native Americans are limited to three studies with inconsistent results although the only study that gives comparative data for other races shown in row 27 gives the highest percentage with experience of sexual intercourse. It should be noted that the race differences in sexual precocity may be explicable in terms of differences in sexual maturation documented by Rushton (2000).

Studies providing data on race differences in sexual precocity assessed as age of first intercourse are summarized in Table 2.37. These studies show that Blacks have earlier first sexual intercourse than Whites and Hispanics, while Asians have the latest first sexual intercourse. Row 12 gives the most recent data from the National Longitudinal Study of Youth.

Table 2.37. Race differences in age of first intercourse

	SEX	ASIAN	BLACK	HISP.	WHITE	REFERENCE
1	M	-	14.4	-	15.9	Zelnik & Shah, 1983
2	F	-	15.5	-	16.4	Zelnik & Shah, 1983
3	MF	16.4	14.4	15.3	16.2	Moore & Erickson, 1985

	SEX	ASIAN	BLACK	HISP.	WHITE	REFERENCE
4	F	-	16.8	-	17.4	Wyatt, 1991
5	M	-	13.4	-	-	Tucker, 1991
6	M	-	14.3	15.4	16.3	Day, 1992
7	F	-	16.8	17.8	17.4	Day, 1992
8	MF	-	15.1	-	-	Lawrence et al., 1995
9	M	17.9	-	-	16.9	Meston et al., 1996
10	F	18.2	-	-	17.0	Meston et al., 1996
11	M	-	13.6	-	-	Samuels, 1997
12	MF	-	15.3	-	16.1	Biello et al., 2013

A further measure of sexual precocity is available in teenage birth rates. Race differences in these are given in Table 2.38. The results show that in the three teenage age groups Blacks had the greatest percentages of teenage births followed by Hispanics, while Whites had the lowest percentages.

Table 2.38. Race differences in teenage birth rates (percentages)

AGE	BLACK	HISPANIC	WHITE	REFERENCE
10/14	0.5	0.2	0.05	Guttmacher, 1994
15/17	8.4	6.5	2.3	Guttmacher, 1994
18/19	16.3	14.8	7.2	Guttmacher, 1994

While the studies summarized in this section provide considerable evidence for greater sexual precocity in Blacks than in

Asians, Hispanics and Whites, this may be attributable to the earlier maturation of Blacks documented by Herman-Giddens, Slora and Wasserman (1987) and Rushton (2000).

21. CHILD MALTREATMENT

We consider next the psychopathic personality characteristic described by the American Psychiatric Association as "inability of function as a responsible parent". This inability can be assessed by child maltreatment, abuse and neglect. Data for racial differences in this are summarized in Table 2.39.

Table 2.39. Race differences in child maltreatment, abuse and neglect (percentages)

	CHILD MALTREATMENT	ASIAN	BLACK	HISPANIC	NATIVE AMERICAN	WHITE	REFERENCE
1	Child abuse	-	2.1	-	-	1.2	Hampton et al., 1989
2	Child neglect	-	0.39	0.32	-	0.28	Lauderdale et al., 1980
3	Children adopted	-	19	-	-	2	Bachraach et al., 1992
4	Children fostered	-	-	-	25	0.05	Westermayer, 1977
5	Infant homicide	0.004	0.020		0.020	0.006	Overpeck et al., 1998

	CHILD MALTREATMENT	ASIAN	BLACK	HISPANIC	NATIVE AMERICAN	WHITE	REFERENCE
6	Infant homicide	-	0.025	0.006	-	0.007	Finkelhor, 1997
7	Child homicide	-	0.009	0.002	-	0.002	Finkelhor, 1997
8	Child neglect	0.5	6.0	1.8	4.2	0.7	Ards et al., 2003
9	Child neglect	0.9	6.0	2.3	5.0	0.7	Ards et al., 2003
10	Maltreatment	-	30.0	20.7	-	13.5	Putnam-Hornstein et al., 2013

Row 1 presents the results of a study of child abuse and neglect collected by the American Association for Protecting Children. They report that in the 1980 census Black children constituted approximately 15 percent of the child population, while from 1976 to 1980 approximately 19 percent of children suffering child abuse and neglect were Black. For 1982, 1984 and 1985, the figures were 22.0, 20.8 and 26.8 percent, respectively. Thus, Black children are about 50 percent over-represented in these figures. Hampton et al. (1989) examined this question further by an analysis of the American First (1975) and Second (1985) National Family Violence Surveys. In these surveys severe violence towards children is defined as hitting them with the fist or with some object (other than those commonly employed in corporal punishment), and kicking, biting, and beating them up. It does not include what is designated "minor violence," consisting of slapping or spanking. The percentages of Black and White children subjected to severe violence in the two surveys for 1975-85 combined are shown in row 1 of Table 2.35, showing that almost twice as many Black children as White were subjected to severe violence.

Row 2 gives the results of a study of child abuse and neglect in Texas over the years 1975-1977. All the 36,945 cases validated by the Department of Human Resources were analyzed by racial and ethnic group. The incidence was calculated in relation to the numbers of Blacks, Whites, and Hispanics in the population, and shows the rates of child abuse and neglect were about 40 percent greater among Blacks than among Whites, and about 14 percent greater among Hispanics than among Whites. Row 3 gives results for a further index of irresponsible parenting consisting of the numbers of children given up by single mothers for adoption, and shows this approximately ten times greater by Blacks than by Whites. Row 4 presents data for children taken away from their parents because of neglect or abuse and put into foster homes or placed out for adoption in the state of Minnesota in the 1960s and 1970s and shows that 25 percent of Native American children were fostered or adopted compared with 0.05 percent of Whites.

The most extreme expression of the inability to function as a responsible parent consists of killing a child Racial differences in the homicide of infants in their first year of life are shown in row 5. The authors of this study examined the histories of approximately 35 million babies born between 1983 and 1991. They found that 2,776 of these had been murdered. They were not able to obtain information on the perpetrators of these homicides but they cite studies showing that the great majority of the infant homicides are carried out by the mothers, or the mothers' husbands or partners. They calculated the rate of infant homicides for Blacks, East Asians, Native Americans and Whites, with the results given in row 5 showing that the infant homicide rate of Blacks and Native Americans is approaching four times as great as that of Whites, while the rate among Asians is about two thirds the White rate. Row 6 gives similar data for 1991-1992 for the murder of infants in their first year of life calculated from Uniform Crime Reports, and shows a Black rate of infant homicide about three and a half times greater than that of Whites and Hispanics. Row 7 extends this analysis to infants aged between 1 and 2 years and shows that for this age group the homicide rate has fallen but that Blacks killed their infant children at about four and a half times the rate of Whites and Hispanics. Rows 8 and 9 give race differences in child neglect in Minnesota in 1993 and 1999 showing the highest rate in Blacks, followed by Native Americans, Hispanics, Asians and Whites. Row 10 gives data for 2002-2007 for the percentages of 74,182 children aged 0-5 years in California referred for maltreatment and shows the rate for Blacks more than twice as great as that for Whites, with Hispanics intermediate.

In further studies race differences in irresponsible parenting have been expressed as odds ratios in relation to the rate for Whites set at zero. Data for these are summarized in Table 2.40. Row 1 gives data for the percentages of neglected children for 1988 and shows that the prevalence of child neglect was 60 percent greater among Hispanics and 80 percent greater among Blacks, as compared with Whites. Row 2 gives the percentages of child maltreatment collected by the United States Department of Health and Human Services for 1996 and shows maltreatment about three times greater among Blacks and about one and a half times greater among Hispanics, as compared with Whites. Row 3 gives the percentages of child maltreatment for five states for 2000, and shows maltreatment about twice as great among Blacks and slightly greater among Hispanics, and much lower among Asians as compared with Whites.

Table 2.40. Race differences in child neglect and maltreatment (odds ratios)

	MEASURE	ASIAN	BLACK	HISP.	WHITE	REFERENCE
1	Neglect	-	1.8	1.6	1.0	American ASPC, 1988
2	Maltreatment	-	3.1	1.5	1.0	US Dept Health, 1996
3	Maltreatment	.58	1.9	1.1	1.0	Fluke et al., 2003

22. SELF-ESTEEM

People with high self-esteem have a high opinion of themselves, their families, and their social capacities and abilities. It has been reported that high self-esteem is positively correlated at .23 with psychopathy in a sample of male prison inmates (Cale & Lilienfeld, 2006), and at .43 with Factor 1 psychopathy (superficial charm, poor moral sense) among college students (Falkenbach, Howe & Falki, 2013). It has been shown in numeous studies that have higher self-esteem that Whites. Twelve studies reporting this were cited by Harris & Stokes (1978), and further studies have been reviewed by Cross (1985) and cited by van Laar (2000, p.35): "African-American students have generally been found to have equal or higher self-esteem than white students."

Recent studies of race differences in self-esteem are given in Table 2.41. Row 1 gives results of a study of 6,504 7-12th grade school students showing the percentages with high self-esteem measured by questions like "I have a lot to be proud of," and shows the highest percentage among Blacks followed by Hispanics and Whites, and the lowest percentage among Asians. Row 2 from the same study gives the percentages scoring above 75 percecnt on the self-esteem questionnaire showing the same race differences. Row 3 gives results of a meta-analysis as deviations from Whites set at zero and shows that Blacks have the highest self-esteem followed by Whites, while Hispanics and Native Americans have lower self-esteem than Whites, and Asians have the lowest self-esteem. Row 4 gives results from a later meta-analysis as deviations from Whites set at .zero and shows that Blacks and Hispanics have higher self-esteem than Whites, and Asians have lower self-esteem than Whites. Rows 5 through 10 give results as scores on self-esteem from a study of 8th, 10th and 12th grade school students and shows that Blacks have the highest self-esteem, followed by about equal scores for Hispanics and Whites, while Asians have the lowest self-esteem. Row 11 gives results for female teenagers on a self-esteem questionnaire showing significantly higher scores by Blacks.

Table 2.41. Race differences in self-esteem (percentages, Ors)

	SELF-ESTEEM	ASIAN	BLACK	HISPANIC	NATIVE AMERICAN	WHITE	REFERENCE
1	MF: %	29	52	38	-	40	Bankston & Zhu, 2002
2	MF: %	63	82	70	-	74	Bankston & Zhu, 2002
3	MF: d	-.30	.19	-.09	-.21	.00	Twenge & Crocke, 2002
4	MF: d	-.30	.17	.25	-	.00	Foldes et al., 2008
5	M: grade 8	4.06	4.14	4.00	-	4.14	Bachman et al., 2010
6	F: grade 8	3.80	4.15	3.83	-	3.98	Bachman et al., 2010
7	M: grade 10	3.88	4.31	4.18	-	4.13	Bachman et al., 2010
8	F: grade 10	3.69	4.22	3.99	-	3.88	Bachman et al., 2010
9	M: grade 12	3.90	4.31	4.20	-	4.16	Bachman et al., 2010
10	F: grade 12	3.76	4.31	4.09	-	3.98	Bachman et al., 2010
11	F: teenagers	-	35.2	-	-	32.4	French & Neville, 2013

23. DRUG AND SUBSTANCE ABUSE

Many studies have found that drug use and abuse are associated with psychopathic personality, e.g. "the relationship between psychopathic behavior and drug abuse is strong" (Fals-Stewart, 2005, p.311);

and "studies have reported that 80–85 percent of individuals with ASPD meet criteria for a substance use disorder ... compared to the estimated US population lifetime prevalence rates of 13.5 percent for alcohol use disorders and 6.1 percent for other drug use disorders" (Glenn, Johnson & Raine, 2013, p 427). It has been reported that psychopathic personality was present in 44 percent of samples with substance abuse and drug dependence (Brooner et al., 1993; Compton, Cottler, Abdallah et al., 2000). It has also been found that opiate addicts score high on Eysenck's psychoticism trait, a measure of psychopathic personality (Doherty & Matthews, 1988). Drug use and abuse are a component of Hare's second psychopathic personality factor consisting of a syndrome of nine "social deviance" characteristics that include poor behavior controls and an inability to control the need for immediate gratification at the expense of long term adverse consequences.

There is such a large literature on race differences in substance abuse that it is not possible to cover it comprehensively here but sufficient studies are given to illustrate the race differences in Table 2.42. Row 1 gives death rates for Blacks and Whites from substance abuse for 1989 published by the National Center for Health Statistics, and shows this was more than twice as great among Blacks as among Whites. Row 2 gives differences in the percentages of those with substance abuse in a sample of 19,688 outpatients seen at county mental health clinics in Los Angles between 1983-1988, and shows the highest percentage in Blacks followed by Hispanics and Whites, with the lowest percentage in Asians divided approximately equally between Northeast Asians (Chinese, Japanese and Koreans) and Southeast Asians (Filipinos, Cambodians, Vietnamese, etc). Row 3 gives current elicit drug use rates for Blacks, Native Americans and Whites from the 1999 Household Survey of Drug Use and shows the highest percentage in Native Americans followed by Blacks and Whites. Row 4 shows the results of a study of 7,740 12-17 year olds males admitted to hospital for traumatic injuries and tested positive for drug use. Row 5 gives the same data for females, and both show the highest percentage in Blacks. Row 6 gives data for drug offenses in 2001 with Whites set at 1, showing the highest rates among Blacks followed by Hispanics and Native Americans and lowest rates among Asians.

Row 7 gives data for drug offenses in 2010 with Whites set at 1, showing the highest rates among Blacks followed by Hispanics and the lowest rates among Whites. Rows 8 and 9 show greater alcohol abuse among Native Americans and less alcohol abuse among Asians compared with Whites. Row 10 gives percentages of Blacks and Whites who consumed five or more drinks a day, and shows greater alcohol consumption by Blacks. These results confirm many studies that have found that Blacks and Native Americans have higher rates of alcohol abuse and of disorders caused by alcohol abuse (e.g. liver cirrhosis, death from alcoholism) than Whites, shown by Kerr, Patterson & Greenfield (2009), and confirmed in a review of forty studies by Zapolski et al. (2014).

Rows 11 through 15 give data for the use of cigarettes. Row 11 shows the highest current use of cigarettes by Blacks, followed by Whites and Hispanics, and the lowest use in Asians. Row 12 shows a higher current use of cigarettes among 18 year olds by Native Americans than among Whites. Row 13 shows a lower current use of cigarettes among Asians than among Whites. Row 14 gives 2011 data showing the highest use of cigarettes among Native Americans and the lowest use among Asians. Row 15 gives 1990s data for Californian school students showing a lower use of cigarettes among Asians than among Whites. Rows 16 through 19 give data for the use of cocaine. Row 16 gives lifetime use of cocaine among 18 year olds, showing the percentage of Native American users higher than that of Whites. Rows 17 and 18 give last month use of cocaine among 31 year old men and women showing the percentage of Blacks users higher than that of Whites.

Row 20 gives lifetime use among 18 year olds presented as odds ratios and shows the percentage of Native American users three times higher than that of Whites. Row 21 gives data for the lifetime use of hallucinogens (e.g. LSD) from a 1988-90 study showing the prevalence rate among Native American 18 year olds greater than that of Whites. Rows 22 through 26 give data for the use of inhalants consisting principally of gasoline (petrol), glue, paint, polish remover and lighter fuel. Row 22 gives data for the prevalence rates of 12-17 year old Native Americans who have used inhalants, and of all Americans of the same age entered as Whites reported by Fishburne, Abelson and Cisin

(1980). Row 23 gives data for the race differences in the lifetime use of inhalants among 18 year olds in a 1988-90 study showing prevalence rates of Native American youth were greater than those of Whites. Row 24 confirms this result and also shows the life time prevalence among Hispanics as the same as that of Whites. Row 25 further confirms this result on a study of over 8,000 secondary school students who "sometimes used inhalants," showing greater prevalence among Native American than among Whites but lower prevalence among Blacks. Row 26 gives lifetime use among Californian school students showing greater use among Whites than among Asians.

Rows 27 through 32 give race differences in use of marijuana (cannabis). Row 27 gives data for the lifetime use of marijuana from a from a 1988-90 study showing the prevalence rate among Native American youth greater than that of Whites. Row 28 confirms the greater lifetime use of marijuana from a study of an American sample of 1,512 13-18 year olds, and also shows a high prevalence among Hispanics. Rows 29 and 30 give results for men and women from the American CARDIA (Coronary Artery Risk Development in Young Adults) study of 5,115 young adults for at least one day of marijuana use during the last month. The results show that Black men used marijuana more than White men, but there was no difference in use between Black and White women. Row 31 gives lifetime use among Californian school students showing greater use among Whites than among Asians. Row 32 gives a further study showing a high use of marijuana by Native Americans.

Row 33 gives data for the lifetime use of stimulants (e.g. amphetamines, ecstasy) from a 1988-90 study showing the prevalence rate among Native American 18 year olds greater than that of Whites. Row 34 confirms the greater lifetime use of stimulants by Native Americans from a study of an American sample of 1,512 13-18 year olds, and also shows a higher prevalence among Hispanics than among Whites.

Table 2.42. Race differences in drug and substance abuse (percentages; ORs)

	ABUSE	ASIAN	BLACK	HISPANIC	NATIVE AMERICAN	WHITE	REFERENCE
1	Substance	-	11.4	-	-	4.8	Nat.Cent., 1991
2	Substance	15.4	30.9	28.0	-	25.7	Flaskerud & Hu, 1992
3	Substance	-	7.7	-	10.6	6.6	Amaro et al., 2001
4	Substance	-	32.8	22.2	-	20.0	Marcin et al., 2003
5	Substance	-	17.8	12.8	-	17.1	Marcin et al., 2003
6	Drug offences	0.2	12.5	5.2	1.8	1.0	Taylor, 2005
7	Drug offences	-	2.8	1.6	-	1.0	Diamond et al., 2012
8	Alcohol	-	-	-	55	35	Walls et al., 2013
9	Alcohol	16.3	-	-	-	32.9	Wong et al, 2004
10	Alcohol	-	30.6	-	-	21.6	Dawson, 1998
11	Cigarettes	12	51	16	-	23	Navarro,1999
12	Cigarettes	-	-	-	75	62	Oetting & Beauvais,1990
13	Cigarettes	20	-	-	-	27	Gardner,1994
14	Cigarettes	13	26	20	43	29	Substance Abuse Admin, 2012
15	Cigarettes	25.2	-	-	-	43.9	Wong et al, 2004

	ABUSE	ASIAN	BLACK	HISPANIC	NATIVE AMERICAN	WHITE	REFERENCE
16	Cocaine	-	-	-	14	8	Oetting & Beauvais,1990
17	Cocaine: m	-	8.4	-	-	2.1	Braun et al.,1996
18	Cocaine: w	-	2.1	-	-	1.2	Braun et al.,1996
19	Cocaine:	2.9	-	-	-	4.2	Wong et al, 2004
20	Heroin	-	-	-	3	1.0	Oetting & Beauvais,1990
21	Hallucinogens	-	-	-	18	10	Oetting & Beauvais,1990
22	Inhalants	-	-	-	31	11	Oetting et al., 1980
23	Inhalants	-	-	-	17	10	Oetting & Beauvais,1990
24	Inhalants	-	-	25	31	25	Swain et al.,1997
25	Inhalants	-	3	-	27	8	Carroll, 1977
26	Inhalants	8.1	-	-	-	13.6	Wong et al, 2004
27	Marijuana	-	-	-	67	38	Oetting & Beauvais,1992
28	Marijuana	-	-	64	61	49	Swain et al.,1997
29	Marijuana: m	-	38	-	-	34	Braun et al.,2000
30	Marijuana: w	-	22	-	-	22	Braun et al.,2000
31	Marijuana	10.4	-	-	-	26.5	Wong et al, 2004

	ABUSE	ASIAN	BLACK	HISPANIC	NATIVE AMERICAN	WHITE	REFERENCE
32	Marijuana	-	-	-	27	17	Walls et al., 2013
33	Stimulants	-	-	-	24	13	Oetting & Beauvais,1990
34	Stimulants	-	-	33	28	26	Swain et al.,1997

The general pattern of the results given in Table 2.41 is that drug and substance abuse are high in Blacks and Native Americans, intermediate in Hispanics, lower in Whites and lowest in Asians.

24. ALTRUISM

Altruism is behavior that improves the welfare of another individual or individuals without improving or even reducing the welfare of the altruist. Altruism is the antithesis of the selfishness and lack of social concern of psychopaths and it has been shown that altruism is significantly negatively correlated with primary psychopathy at -.56 and with secondary psychopathy at -.33 (White, 2014). The neurophysiological basis of the altruism-psychopathy dimension is that altruists have an enhanced right amygdala volume and emotional responsiveness to viewing fearful expressions (a measure of empathy), while in psychopaths the right amygdala volume and responsiveness are reduced (Marsh et al, 2014). It would therefore be expected that there would be race differences in altruism consistent with the differences in a number of psychopathic behaviors that have been documented in previous sections. To assess this expectation, we first examine race differences in organ donation as an expression of altruism because it has been shown by Yeung, Kong and Lee (2000), and by Morgan and Miller (2002), that those who are willing to donate organs are highly altruistic.

Studies of race differences in willingness to donate organs in the event of death and while alive, in the case of kidney donation, are summarized in Table 2.43. Row 1 gives percentage rates endorsing the question "I am willing to donate my organs or tissues at the time of my death" in a 1994 study of 683 college students, and shows the percentage significantly lower in Blacks than in Whites, with Asians and Hispanics intermediate. It was also found in this study that there was little difference between the races in knowledge of organ transplantation; the question "I am knowledgeable about organ procurement and about organ procurement system" was endorsed by 36.6 of Blacks, 37.5 percent of Hispanics, 41.7 percent of Asians, and 39.0 percent of Whites. Row 2 gives rates of kidney donation per million population during 1991-1993 and shows higher donation rates by Whites (9.20) than by Blacks (8.85). This study also found that women had higher rates of donation than men among both Blacks (10.0 vs 7.7) and Whites (10.3 vs 8.1), indicating greater altruism in women, consistent with many studies showing that women are more altruistic and less psychopathic than men. Row 3 gives results from a study of Koreans in New York showing a high percentage of willing organ donors. Row 4 gives results from a study of over 6,000 telephone interviews reporting that 43 percent of Whites, but only 31 percent of Hispanics and only 23 percent of Blacks were willing to donate their organs after death. Row 5 gives results from a study of 278 ethnic Vietnamese in Seattle, showing a high percentage (51.3 percent) were willing to donate organs after death. Row 6 gives results from a survey of high school students to the question 'I would like to become an organ donor,' and shows more positive responses by Whites (39 percent) than by Blacks (24 percent). Row 7 gives results from a survey of college students again showing a much higher percentage of Whites than of Blacks willing to donate organs. Row 8 gives race differences expressed as odds ratios with the rate for Whites set at 1.0 in a study of 883 students at nine inner-city high schools in Seattle who were asked whether they had signed an organ donor card. The results show that Whites were more willing to donate than minorities, such that in relation to one White donor there were 0.42

Black donors, 0.34 Asian donors and 0.40 Hispanic donors. This study also found that girls were significantly more willing to donate than boys.

Table 2.43. Race differences in willingness to donate organs (percntages)

	ASIAN	BLACK	HISPANIC	WHITE	REFERENCE
1	39.6	36.6	45.3	59.2	Rubens, 1996
2	-	8.85	-	9.20	Bloembergen et al., 1996
3	54.5	-	-	-	Joun et al., 1997
4	-	23	31	43	McNamara et al., 1999
5	51.3	-	-	-	Pham & Spigner, 2004
6	-	24	-	39	Spigner et al., 2002
7	-	35	-	67	CORT & CORT , 2008
8	.34	.42	.40	1.0	Thornton et al, 2006

For a second expression of altruism we examine charitable giving. Several studies have examined race differences in this{charitable giving/race differences in}, and have found that Asians, Blacks, and Hispanics, donate less to charities than do Whites, controlling for income disparities. Brown & Ferris (2007) found that, compared to Whites, Hispanics donate significantly less to religious purposes, and both African Americans and Hispanics donate less to secular purposes. Wang and Graddy (2008) in reviewing the research write that, "Studies on race/ethnicity differences in charitable giving have consistently shown that being white is positively associated with the likelihood to donate or to donate higher amounts compared to other minority groups. Surveys have revealed that African Americans and Hispanics have lower rates of household giving and gave less amounts compared to whites," and they published additional data confirming these race differences.

Further evidence for race differences in charitable giving has been reported by Leslie, Snyder, and Glomb (2013) in a study in which they investigated race differences in charitable donations to a workplace charity in a sample of 16,429 employees in a university. Most participants were non-academic staff (78 percent) rather than faculty. They gathered data on the amount each employee donated during the organization's annual month-long charitable giving campaign. Employees were invited to donate to charities concerned with alleviating poverty, the support of education, and the treatment of illness. In examining the race differences in the amount of charitable giving, they controlled for salary, position, and age, because these are likely to affect the amount of charitable donations. They found that women donated $31 more to the workplace charity drive than did men, consistent with other research showing that women are more altruistic than men. They also found that all the minorities donated less to charity than did Whites, by an average $26. However, Blacks gave more than Whites to the Black charity. Their results show that Whites are more altruistic than other races in giving more to charities that support all races, while Blacks support their own race by charitable giving but give less support than do Whites to other races.

For a third expression of race differences in altruism we examine a study of "organizational citizenship behavior" (OCB) defined as "individual behavior that is discretionary, not directly or explicitly recognized by the formal reward system, and going beyond the call of duty." Actions that are examples of OCB include such things as assisting others and volunteering for activities not related to an individual's job description, for example, planning the company picnic. It has been reported that Whites show greater organizational citizenship behavior than do Blacks (Jones & Schaubroeck, 2004).

25. EMOTIONAL INTELLIGENCE

Emotional intelligence (EI) is defined by Coleman (2008) as the ability to monitor one's own and other people's emotions, to discriminate between different emotions and label them appropriately, and to use emotional information to guide thinking and behavior. This ability is weak in psychopathic personality, and race differences in it would be expected. This has been confirmed by Whitman, Kraus, and van Rooy (2014) in a study of 209 Blacks and 125 Whites applying to become fire fighters tested with the 16-item Wong and Law Emotional Intelligence Scale. Whites scored 6.41 (Sd 0.51)—higher than Blacks (6.14 (Sd 0.81) by approximately a third of a standard deviation (d = .32).

26. RACE DIFFERENCES IN CRIMINAL SAMPLES

There have been three meta-analyses that have reported that in criminal samples Blacks have higher levels of psychopathic personality than Whites. In the first, Skeem, Edens, Sanford, and Colwell (2003) have shown that in nine prison samples blacks had higher psychopathic personality by an average of $0.14d$. In the second study, Skeem, Edens, Camp, & Colwell (2004) reported that in 21 studies of correctional, substance abuse, and psychiatric samples, that Blacks were more psychopathic than Whites by $0.11d$, a statistically significant difference. In the third study, McCoy and Edens (2006) reported that in 16 studies of youths with criminal records, Blacks were more psychopathic than Whites by $0.20d$, a statistically significant difference. Skeem, Edens, Sanford, and Colwell (2003) argue that the difference is negligible and that therefore that there is no significant difference between Blacks and Whites in psychopathy. This is a misunderstanding because most of those in criminal samples have already been selected for high psychopathic personality and hence little difference between Blacks and Whites would be expected. Analogously, although there is a large difference between Blacks and Whites in average intelligence, a large difference between Blacks and

Whites in average intelligence would not be expected in samples of the mentally retarded. The small but consistently higher and statistically significant rates psychopathic personality in Blacks than in Whites, reported in the three meta-analyses, is further testimony to its higher rate present in numerous general population samples.

Further studies of psychopathic personality in prison samples are summarized in Table 2.44. Row 1 gives results for a study showing that among prisoners, 57 percent of Blacks and 43 percent of Whites received misconduct reports. Row 2 gives results for violent prison misconduct expressed as odds ratios with Whites set at zero, in which Hispanics showed the greatest violent misconduct followed by Blacks, while Whites and "others," largely Asians and entered as such, showed the least.

Table 2.44. Race differences in psychopathic behavior in prison samples

	MEASURE	ASIANS	BLACKS	HISP.	WHITES	REFERENCE
1	Misconduct: %	-	57	-	43	Ramirez, 1983
2	Violence: OR	.00	.09	.13	.00	Diamond et al., 2012

27. RACE DIFFERENCES IN THE FIVE FACTOR PERSONALITY MODEL

In recent years the major theory of personality is the five-factor model consisting of anxiety, introversion-extraversion, conscientiousness, agreeableness, and openness to experience. In terms of this model, it has been shown that psychopathic personality is associated most strongly with low conscientiousness (Widiger & Lynam, 1998; De Cuyper, De Fruyt & Buschman, 2008). It has also been shown that academic dishonesty is correlated at -.22 with conscientiousness (Giluk & Postlethwaite, 2015). Race differences in conscientiousness in a sample of 3629 working adults have been reported by Goldberg, Sweeney, Merenda, and Hughes (1998), and are given as correlations with Whites such that negative correlations denote lower scores than those of Whites. Blacks and Hispanics obtained significantly negative correlations of -.24 for Blacks and the -.23 for Hispanics. Asians and Native Americans also obtained negative correlations of -.10 and -.07 but these were not statistically significant.

28. CONCLUSIONS

The pattern of results for the numerous measures summarized in this chapter is that psychopathic personality is greatest in Blacks and Native Americans, followed by Hispanics, lower in Whites, and lowest in Asians, especially in Northeast Asians where data are given for these disaggregated from Southeast Asians.

CHAPTER 3

CANADA

The adult population of Canada consists of approximately 90 percent of Europeans, two percent ethnic Chinese and three percent of Aboriginals (also designated First Nations), of whom approximately three quarters are Native American Indians and a quarter are Inuit. The remaining 5 percent are unclassified. Among those aged under 16 approximately 5 percent are Aboriginals.

1. CRIME

Race differences in rates of crime for males and females combined are given in Table 3.1. Row 1 gives crime rates of juvenile delinquents in Vancouver in 1928/36, showing rates of Whites 15.6 times higher than those of ethnic Chinese and Japanese. Row 2 gives rates of admission to prison for 1986 showing that these were approximately eight times as great for Native Americans as for the rest of the population. Row 3 gives race differences in convictions for violent crime per 1,000 for the Canadian province of Ontario and consists of prison admissions for 1992, showing admissions of Blacks about five times greater than those of Whites and admissions of Native Americans about two and a half times those of Whites, while admissions for East Asians were about half those for South Asians and about two thirds those of Whites. These admission figures are snapshots for those in prison on a particular date of the year. Foran (1995) reported that in the mid-1990s Aboriginal adults were 3 percent of the population and 17 percent of men and 26 percent of women in prison.

Table 3.1. Race differences in crime in Canada, per 1,000

CRIME	BLACK	EAST ASIAN	NATIVE AMERICAN	SOUTH ASIAN	WHITE	REFERENCE
All crime	-	1.0	-	-	15.6	MacGill, 1938
Imprisonment	-	-	26.0	-	3.0	Tonry, 1994
Violent	36.9	3.5	19.9	4.6	7.1	Ontario, 1996

2. LONG-TERM MONOGAMOUS RELATIONSHIPS

Race differences in the percentage married as a measure of the capacity to form long-term monogamous pair bonds with members of the opposite sex are given for 36 year olds for 1996/97 for men and women combined, in Table 3.2. It will be seen that Whites had the highest percentage married followed in descending order by Asians and Hispanics, Native Americans, and Blacks.

Table 3.2. Race differences in percentages married

BLACK	ASIAN	HISPANIC	NAT.AMER.	WHITE	REFERENCE
43	54	52	45	60	Wu et al., 2003

3. INTIMATE PARTNER VIOLENCE

Studies of intimate partner violence by men against their women partners are summarized in Table 3.3. Row 1 gives results for the percentage of men who had forced women to have sex against their will showing the highest percentage among Blacks, followed by South Asians and the lowest percentage in Hispanics. No results for Whites were given in this study. Row 2 shows three times as many Native American women experiencing violence from their partners during the last five years. Row 3 gives numbers of women killed by their partners during 1973-2000 in Northwest Territories and Yukon, largely inhabited by Native Americans, and shows these approximately seven times greater that the rates for Canada as a whole.

Table 3.3. Women with experience of intimate partner violence

	INTIMATE PARTNER VIOLENCE	BLACK	HISPANIC	NATIVE AMERICAN	SOUTH ASIAN	WHITE	REFERENCE
1	Lifetime percent	19	1	-	5	-	Maticka-Tyndale et al., 1996
2	Last 5 years percent	-	-	25	-	8	Health Canada, 1997
3	Homicide per million	-	-	77.8	-	11.5	Dookie, 2004

4. RECKLESSNESS

Studies of race differences in recklessness measured by the non-use or inconsistent use of contraception are summarized in Table 3.4. Row 1 gives data for the percentages of men who were in long-term relationships who did not use contraception showing the lowest percentage among South Asians. Row 2 gives results for the inconsistent use of contraception by men with new sexual partners in the past year, showing the highest percentage among Blacks, followed by South Asians, and the lowest percentage in Hispanics No data for Whites were given in this study.

Table 3.4. Race differences in recklessness (percentages)

	USE CONTRACEPTION	BLACK	HISPANIC	SOUTH ASIAN	REFERENCE
1	No use	23	22	12	Maticka-Tyndale et al., 1996
2	Inconsistent	70	57	66	Maticka-Tyndale et al., 1996

5. MULTIPLE SEXUAL PARTNERS

A study of race differences for men in multiple sexual partners are given in Table 3.5, showing that having two or more sexual partners in the past year was higher among Blacks than among Asians. No data for Whites were given in this study.

Table 3.5. Race differences in multiple sexual partners (percentages)

N PARTNERS	BLACK	ASIAN	REFERENCE
Past year 2 +	55	31	Maticka-Tyndale et al., 1996

6. IRRESPONSIBLE PARENTING

Irresponsible parenting measured by child maltreatment, abuse and neglect is considerably more prevalent in Native American and Aboriginal including Inuit children in Canada than among Europeans,. In a review of the literature, Fluke, Chabot and Fallon (2013, p. 47) have written, "The chronic over-representation of Aboriginal children in Canadian child welfare care has been well documented. Analysis based on national census data noted that while 5 percent of children in Canada were Aboriginal in 1998, Aboriginal children made up 17 percent of children reported to the child welfare, 22 percent of substantiated reports of child maltreatment, and 25 percent of children placed in care in Canada."

7. DRUG AND SUBSTANCE ABUSE

The Department of Indian Affairs and Northern Development (2003) reported that 62 percent of First Nations people, aged 15 years and older had an alcohol abuse problem and 48 percent had a drug abuse problem. Studies summarizing rates of drug and substance abuse in Native Americans and Europeans are given in Table 3.6. Row 1 shows higher rates of marijuana use among 15 year olds by Native Americans than by Europeans. Rows 2, 3 and 4 show high rates of 12-month use of marijuana, cocaine, and hallucinogens in a study of 281 Aboriginal peoples aged 35 years in a Canadian city. The author does not give the percentages of Europeans, but in a review of research she writes: "Illicit and prescription drug use disorders are two to four times more prevalent among Aboriginal peoples in North America than in the general population." (Currie, 2013)

Table 3.6. Race differences in drug and substance abuse (percentages)

	DRUG	NATIVE AMERICANS	EUROPEANS	REFERENCE
1	Marijuana	26.7	14.1	Adlaf et al., 2004
2	Marijuana	56.1	-	Currie, 2013
3	Cocaine	32.5	-	Currie, 2013
4	Hallucinogens	14.6	-	Currie, 2013

8. CONCLUSIONS

The pattern of results summarized in this chapter is that race differences in rates of psychopathic personality in Canada are similar to those in the United States, being greatest in Blacks and Native Americans, lower in Whites, and lowest in Asians.

CHAPTER 4

EUROPE

Studies of race differences in the prevalence of psychopathic personality disorder in Europe are available for Black immigrants from the Caribbean and Sub-Saharan Africa, for South Asians from the Indian Subcontinent, and for East Asians consisting mainly of Chinese origin, principally from Hong Kong and Malaysia.

1. PREVALENCE STUDIES

Studies of the prevalence of psychopathic personality in the population in Britain and Norway are summarized in Table 4.1.

Table 4.1. Prevalence of psychopathic personality in Europe (percentages, odds ratio)

	COUNTRY	BLACK	S. ASIAN	WHITE	REFERENCE
1	Britain	2.28	-	0.97	Coid et al., 2009
2	Britain	1.44	1.14	1.0	Crawford et al., 2011
3	Norway	-	-	0.7	Torgersen et al, 2001

Row 1 gives the results from a national survey of a representative sample in Britain, and shows the prevalence of psychopathic personality in Blacks more than twice as great as in Whites. Row 2 gives British results for a larger survey of the prevalence of psychopathic personality in a representative national sample of 8,351, expressed as odds ratios with Whites set at 1.0, and shows the prevalence greatest in Blacks, lower in South Asians, and lowest in Whites. Row 3 gives the results from a national survey of a representative sample of 18 to 65 year olds in Norway and shows the prevalence of 0.7 percent, closely similar to that of 0.97 of Whites in Britain.

2. CONDUCT DISORDERS

Studies reporting race differences in conduct disorders and the associated behaviors of delinquency and impulsivity are summarized in Table 4.2. Row 1 gives data for Britain and shows the percentages of conduct disorder substantially higher among Blacks than among Whites. Rows 2 and 3 give results of a study of 13 year olds in Britain, showing Blacks more delinquent and more impulsive than Whites. Row 4 gives data for Britain for 7 year olds showing greater conduct disorders in Blacks as the average of Black Africans and Black Caribbeans, approximately the same rate of conduct disorders in South Asians as the average of Indians, Pakistanis and Bangladeshis, and a lower rate of conduct disorders in Chinese. Row 5 gives further data for Britain for 11 year olds showing greater conduct disorders in Blacks as the average of Black Africans and Black Caribbeans, slightly lower conduct disorders in South Asians as the average of Indians, Pakistanis and Bangladeshis, and lower conduct disorders in Chinese. Row 6 gives results in the Netherlands from a study in which 12-17 year old boys reported on their own delinquent behaviors of stealing, fighting, vandalism, etc. and shows the prevalence of these delinquent behaviors about twice as great among Blacks from the former Dutch colony of Suriname and as among Whites, together with the percentage among South Asians (originally from the Indian Subcontinent) that is only marginally higher than that of Whites. Row 7 shows further results in the Netherlands from a study of Turkish 11-18 year old immigrants who displayed greater externalizing behavior problems and delinquency than indigenous Whites.

Table 4.2. Race differences in conduct disorders (percentages, ds)

	COUNTRY	DISORDERS	BLACK	CHINESE	SOUTH ASIAN	WHITE	REFERENCE
1	Britain	Conduct: %	38	-	-	10	Rutter et al., 1974
2	Britain	Delinquency: d	.12	-	-	00	Lynam et al., 1993
3	Britain	Impulsivity: d	.20	-	-	00	Lynam et al., 1993
4	Britain	Conduct: d	.25	-.35	.02	00	Lynn & Cheng, 2016
5	Britain	Conduct: d	.22	-.61	-.13	00	Lynn & Cheng, 2016
6	Netherlands	Conduct: %	33	-	17	15	Junger & Polder, 1993
7	Netherlands	Externalising:%	-	-	.32	.26	Van Oort et al., 2007
8	Netherlands	Externalising:%	-	-	.22	.19	Van Oort et al., 2007
9	Netherlands	Delinquency: %	-	-	.26	.19	Van Oort et al., 2007
10	Netherlands	Delinquency: %	-	-	.13	.10	Van Oort et al., 2007

Row 8 gives similar results for Turkish 21-28 year olds. Rows 9 and 10 show Turkish 11-18 year old and 21-28 year olds immigrants in the Netherlands with more in self-reported delinquencies than indigenous Whites.

Further studies have reported race differences in conduct disorders in Britain as odds ratios (ORs) giving the percentages of conduct disorders with the percentage of whites set at 1.0. These are summarized in Table 4.3.

Table 4.3. Race differences in conduct disorders in children (odds ratios)

	SEX	BLACK	EAST ASIAN	SOUTH ASIAN	WHITE	REFERENCE
1	M	3.9	-	-	1.0	Tizard et al., 1988
2	F	2.3	-	-	1.0	Tizard et al., 1988
3	M/F	1.4	-	-	1.0	Goodman & Richards, 1995
4	M/F	4.4	0.18	0.92	1.0	Gillborn & Gipps, 1996
5	M/F	2.2	0.93	2.59	1.0	Tippett et al., 2013
6	M	1.07	-	.50	1.0	Gonzales et al., 2014

Row 1 gives results for 292 Black and 1311 White children and adolescents referred for psychiatric problems to the Maudsley Hospital in London and show the proportion with conduct disorders among the Blacks boys was 3.9 times greater than among Whites, and among Black girls 2.3 times greater than among Whites (Whites had proportionately more emotional disorders). Row 3 gives behavior problems assessed by teachers of Black and White 3-5 year-olds and shows that Black boys and girls had 1.4 times the scores of White children. Row 4 gives rates of conduct disorders for four racial groups and shows these are lowest for East Asians (largely Chinese), next lowest for South Asians, followed by whites, and greatest for Blacks.

Row 5 gives results for racial differences in bullying defined as aggressive behavior, engaged in repeatedly, by an individual or group of peers with more, actual or perceived, power than the victim. The study assessed racial differences in bullying in a British sample of 4,668 adolescents, assessed by self-report answers to questions such as "Do you physically bully other children at school by hitting or pushing them around, threatening or stealing their things?" The results show that bullying is greatest in South Asians and Blacks, lower in Whites, and lowest for East Asians (largely Chinese). Row 6 gives data for violence in Britain assessed by participation in a fight during the last five years. The results are expressed as odds raitos with scores for Whites set at 1.0, and show the rate for Blacks higher than for Whites and the rate for South Asians lower than that for Whites.

3. SCHOOL SUSPENSIONS AND EXPULSIONS

Children are suspended or excluded from schools because of their constant conduct disorders. Exclusions in Britain can be either temporary designated "fixed term" or permanent. In England head teachers have the right to exclude children where "allowing the child to remain in school would be seriously detrimental to the education or welfare of the pupil, or that of others at the school" (Gillborn & Gipps, 1996, p. 52). The principal reasons for exclusions are "disobedience in various forms – constantly refusing to comply with school rules, verbal abuse or insolence to teachers." (Gillborn & Gipps, 1996, p. 53). Fixed term exclusions are more commonly used than permanent exclusions.

Race differences in school suspensions and exclusions in England are given in Table 4.4 from data collected by the British Government Department for Education and Employment that have been analyzed for all exclusions, permanent and fixed term, for the school year 1993/1994 by Gillborn & Gipps (1996). The percentages of these groups excluded from secondary schools were calculated for approximately 30 percent of English secondary schools and based on approximately 1 million children. The results show that Black children were excluded about four times as frequently as White, South Asian children were excluded a little less frequently than Whites at 2.5 percent compared with 2.7 percent, while East Asians (Chinese) had much lower rates at 0.5 compared with 2.7 percent for Whites.

Table 4.4. Race differences in school suspensions and exclusions in England (percentages)

BLACK	EAST ASIAN	SOUTH ASIAN	WHITE	REFERENCE
11.0	0.5	2.5	2.7	Gillborn & Gripps, 1996

4. CRIME

Race differences in crime are given for England, the Netherlands and France in Table 4.5. Row 1 presents imprisonment rates for England for 1993 for 16-19 year-old men, and shows that the rate for Blacks is 5.9 greater than that for Whites, while the rate for South Asians is about two thirds that of Whites. Row 2 gives the imprisonment rates in England for 1993 for 20-39 year-old men for Blacks, Whites, and South Asians, and shows that the incarceration rate for Blacks is 6.4 times greater than that for Whites and South Asians. Row 3 gives data for France for 1995 showing imprisonment rate of Blacks about eight times greater than that of Whites. Row 4 gives data for the Netherlands for 2003 showing arrest rates of Blacks about five times greater than those of Whites, South Asians about three times greater than those of Whites, and Northeast Asians slightly lower than those of Whites. This study also gave rates for the Netherlands Antilles (810), Moroccans (570), Latin America (300), Southeast Asia (200), and Europe (190).

Table 4.5. Race differences in imprisonment and arrests for crime per 10,000 population

	COUNTRY	SEX	BLACK	NE ASIAN	S ASIAN	WHITE	REFERENCE
1	England	M	203.9	-	25.8	34.5	Smith, 1997
2	England	M	211.8	-	32.9	33.1	Smith, 1997
3	France	M/F	54.6	-	-	6.7	Tournier, 1997
4	Netherlands	M/F	570	110	353	120	Blom & Jennissen, 2013

Race differences in convictions for crime presented as odds ratios are given in Table 4.6. Rows 1 and 2 give age-adjusted rates of imprisonment in Britain for 1995 expressed as odds ratios, with the rate for the total population set at 1.00. The odds ratio for Whites at 0.88 is slightly lower than the national average for both men and women. The odds ratio for Black men is far greater at 7.1 while for Black women it is even greater at 12.2. The principal reason for the very high percentage of Black women in prison is that they are used as "mules" by Black men for smuggling drugs into Britain. A significant number of these women are detected and sentenced to terms of imprisonment. South Asian men are in prison at the same rate as White men, but South Asian women are in prison at a higher rate than white women. Northeast Asian (Chinese) men and women are both substantially under-represented in prison. Row 3 presents data for 15-18 year old men in prison in England in 2011 and shows Blacks approximately ten times over-represented, South Asians approximately forty percent over-represented, and Northeast Asians approximately ten percent under-represented. Row 4 presents data for Norway for 2002 showing convictions for robbery by Blacks (Somalis) are 15 times greater than for Whites, and for South Asians (Pakistanis) 5.7 times greater than for Whites. Row 5 presents data for Sweden for 1985/9, and shows the imprisonment of Blacks about two and a half times greater than that of Whites and Northeast Asians, although unusually the crime rate of Northeast Asians is fractionally higher than that of whites.

Table 4.6. Race differences in convictions for crime (odds ratios)

	COUNTRY	SEX	BLACK	NE ASIAN	S ASIAN	WHITE	REFERENCE
1	Britain	M	7.1	0.66	0.87	0.88	Home Office, 1998
2	Britain	F	12.2	0.66	1.10	0.88	Home Office, 1998
3	England	M	7.8	0.54	1.1	0.67	Summerfield, 2011
4	Norway	M/F	15.0	-	5.7	1.0	Kolsrud, 2002
5	Sweden	M/F	2.4	1.10	-	1.0	Martens, 1997

A number of cases of multiple rapes by gangs occurred in England in 2012 and an inquiry into this was made by the Children's Commissioner (2012) for England that contained statistics of the percentages of Blacks, South Asians and Whites convicted of these offences. These were Blacks: 17 percent of convictions, South Asians: 33 percent of convictions; mixed race: 1.8 percent of convictions; and Whites: 43 percent of convictions. These statistics are calculated as percentages over or under-representations of the four groups in the convictions in relation to their percentages in the population given in the 2011 census, using the figures of 2.8 percent of the population for Blacks, 6.7 percent of the population for South Asians, 3.8 percent of the population for mixed race, and 88 percent of the population for Whites. The results are given in Table 4.7 and show that Blacks are over-represented by a factor of 6.07, South Asians by a factor of 4.93, mixed race by a factor of 2.11, and Whites by a factor of 0.49, i.e. less than 1.0 and therefore under-represented. The convicted South Asians were almost entirely of ethnic Pakistani Muslim origin, and not of Hindus or Sikhs.

Table 4.7. Race differences in multiple rapes by gangs in England in 2012

BLACK	SOUTH ASIAN	MIXED	WHITE	REFERENCE
6.07	4.92	2.11	0.49	Children's Commissioner, 2012

As in the United States, some social scientists in Europe have suggested that the race differences in crime conviction rates are caused or exacerbated by prejudice, ethnic bias and racism in the police and the courts. For instance, in Britain Rutter, Giller, and Hagell (1998) have reviewed the evidence on race differences in convictions and conclude: "there are substantial differences in the rates of crime among ethnic groups. These differences are exaggerated by small (but cumulative) biases in the ways in which judicial processing takes place...." (p. 246). They do not give any evidence for this assertion and the evidence is against it. A study in Britain concluded that 80 percent of Black-White incarceration differences "can be accounted for by the greater number of Black offenders who appeared for sentence... and by the nature and circumstances of the crimes they were convicted of." (Tonry, 1994, p. 108). The evidence for this contention has also been reviewed by Smith (1997), who concludes that "in large part the difference in rates of arrest and imprisonment between Black and White people arises from a difference in the rates of offending." If racism is present, it is difficult to understand why the conviction rates of Indians and Pakistanis are about the same as those of Europeans, and Chinese rates are lower than those of Europeans, because many Europeans are prejudiced against these. Furthermore, these race differences are also present in self-reported delinquencies in the study in the Netherlands shown in Table 4.2.

5. MONOGAMOUS RELATIONSHIPS

The propensity to form monogamous relationships based on love can be measured by the extent to which people enter into marriage or stable co-habitation. Race differences in these are shown in Table 4.8.

Table 4.8. Race differences in marriage and co-habitation (percentages)

	COUNTRY	SEX	BLACK	EAST ASIAN	SOUTH ASIAN	WHITE	REFERENCE
1	Britain	M	48	73	88	65	Berrington, 1996
2	Britain	F	39	81	88	72	Berrington, 1996
3	Britain	M/F	38	-	77	68	Modood & Berthoud, 1997
4	Britain	M/F	69	-	97	93	Modood & Berthoud, 1997
5	France	M	45	-	-	55	Model et al., 1999
6	France	F	40	-	-	56	Model et al., 1999

A further measure of racial differences in stable monogamous relationships can be obtained from a study of the percentages of adolescents who lived with both natural parents in a British sample of 4668 adolescents. The results are given in Table 4.9, and show that East Asians (largely Chinese) had the highest percentage living with both natural parents followed by South Asians and Whites, while Black had the lowest.

Table 4.9. Race differences in living with both natural parents in Britain (percentages)

BLACK	EAST ASIAN	SOUTH ASIAN	WHITE	REFERENCE
28.3	50.0	54.4	40.4	Tippett et al., 2013

6. MULTIPLE SEXUAL PARTNERS

Racial differences in the propensity to form long-term monogamous relationships are also expressed in the numbers of multiple sexual partners. Data for these for Britain are summarized in Table 4.10.

Table 4.10. Race differences in multiple sexual partners in Britain (percentages)

	MEASURE	SEX	BLACK	SOUTH ASIAN	WHITE	REFERENCE
1	Last 5 years: 2+	M	46	28	35	Johnson et al., 1994
2	Last 5 years: 2+	F	28	8	23	Johnson et al., 1994
3	Last 2 years: 2+	M	28	18	22	Johnson et al., 1994
4	Last 2 years: 2+	F	13	6	12	Johnson et al., 1994

Rows 1 through 4 give data from a study of a nationally representative sample of approximately 20,000 16-59 year olds carried out in 1990. Rows 1 and 2 give the male and female percentages of Blacks, South Asians, and Whites, who had had two or more sexual partners during the last five years and rows 3 and 4 give the male and female percentages who had had two or more sexual partners during the last two years. For all four comparisons Blacks had more sexual partners, and South Asians had fewer partners than Whites.

7. RECKLESSNESS

Recklessness is a component of psychopathic personality and is defined as a disregard of one's own or others' personal safety. Measures of recklessness can be obtained from a variety of sexual behaviors. We consider first the non-use of contraception by those who do not wish to have children and is reckless both because it is likely to result in an unwanted pregnancy and also because it incurs the risk of contracting sexually transmitted diseases, including HIV and AIDS. Studies of racial and ethnic differences in the non-use of contraception in Britain are summarized in Table 4.11.

Table 4.11. Race differences in the non-use of contraception (percentages)

	MEASURE/ AGE	SEX	BLACK	SOUTH ASIAN	WHITE	REFERENCE
1	Last year/16-59	M	10	4	6	Johnson et al.,1994
2	Last year/16-59	F	5	2	4	Johnson et al.,1994
3	Teen births	F	21	6	6	Modood & Berthoud,1997

Rows 1 and 2 give data from a 1994 survey of approximately 20,000 16-59 year olds, and give the percentages of males and females who had practiced "unsafe sex" during the last year, defined as sexual relations without the use of contraception by those not wishing to achieve a pregnancy. The results show that for both men and women Blacks had the highest percentages followed by Whites, while South Asians had the lowest percentages. Row 3 shows the results of a survey of teenage births carried out in 1994. Virtually no teenage births are planned so they can be regarded as a result of the reckless non-use of contraception. It will be seen that teenage births were three and a half times more prevalent among Blacks than among Whites and South Asians.

Another expression of recklessness is problem gambling. Race differences of this in Britain are summarized in Table 4.12. Row 1 shows rates of problem gambling in a national survey highest in South Asians, intermediate in Blacks and lowest in Whites. Rows 2 and 3 confirm substantially higher rates in South Asians than in Whites (data for Blacks were not reported).

Table 4.12. Race differences in Britain in problem gambling (percentages)

	GROUP	BLACK	SOUTH ASIAN	WHITE	REFERENCE
1	Adults	2.4	3.1	1.8	Ipsos Mori, 2009
2	Men	-	3.36	1.56	Forrest & Wardle, 2011
3	Women	-	5.59	0.29	Forrest & Wardle, 2011

8. SEXUAL PRECOCITY

Studies providing data on race differences in sexual precocity as an expression of psychopathic personality in young adolescents are summarized in Table 4.13. Rows 1 and 2 give results from a British 1990 study of the percentages of males and females who had had sexual intercourse by the age of fifteen years, and shows this was highest for Blacks, intermediate for Whites, and lowest in South Asians. The much greater sexual experiences of males than of females in the South Asian samples seems anomalous, but is explained partly by the much greater sexual experiences with prostitutes of South Asian males, 18 percent as compared with six percent of Whites, and seven percent of Blacks.

Table 4.13. Race differences in precocious sexuality by age of fifteen years (percentages)

	SEX	BLACK	SOUTH ASIAN	WHITE	REFERENCE
1	M	26	11	19	Wellings et al., 1994
2	F	10	1	8	Wellings et al., 1994

9. INTIMATE PARTNER VIOLENCE

A study of lifetime intimate partner violence by men against their women partners in Spain has reported that 27.3 percent of North African (Moroccan) women had experienced violence compared with 14.3 percent of indigenous Spaniards (Vives-Cases et al., 2009).

10. CHILD MALTREATMENT

A study of race differences in child maltreatment in three studies in the Netherlands is summarized in Table 4.14. The results are expressed as differences in over-representation or under-representation of what would be expected if all races had equal rates of child maltreatment. Row 1 gives data assessed by "Sentinels" (professional social workers), and shows that Whites (indigenous Dutch) are under-represented at 86 percent, South Asians (mainly Turks), and North Africans (mainly Moroccans), are over-represented at 2.95 percent and Blacks are over-represented at 4.94 percent. Row 2 gives similar differences from a CPS (Child Protective Services) study. Row 3 gives similar, although smaller, differences from a self-report study showing that the higher rates of child maltreatment by South Asians/North Africans and Blacks are not wholly attributable to prejudice by indigenous professionals. The smaller differences in the self-report study could be attributable to prejudice by indigenous professionals or to immigrants under-reporting the extent of child maltreatment.

Table 4.14. Race differences in child maltreatment

	MEASURE	BLACK	SOUTHASIAN/ NORTH AFRICAN	WHITE	REFERENCE
1	Sentinels	4.94	2.95	0.86	Alink et al., 2013
2	CPS	5.71	4.49	0.76	Alink et al., 2013
3	Self report	2.42	1.65	0.91	Alink et al., 2013

11. DRUG AND SUBSTANCE ABUSE

Race differences in drug abuse are shown in Table 4.15. Row 1 gives data for men and women combined assessed by the percentages aged 15+ with alcoholic psychosis in the Netherlands, and shows this was highest for Blacks, intermediate for Whites, and lowest in South Asians. Rows 2 through 9 give results from a self-report school-based survey of drug taking during the last year among 15 and 16 year olds in England. Rows 2 and 3 show marijuana use highest for Blacks, intermediate in Whites, and lowest in South Asians. Rows 4 though 7 show inconsistent results for ecstasy and stimulants/hallucinogens. Rows 8 and 9 show use of opiates highest in South Asians, intermediate in Blacks, and lowest in Whites. Rows 10 through 15 give results for a representative sample of 13-14 years old in London showing marijuana use in the last month highest for Blacks, intermediate for Whites and lowest in South Asians, confirming the results in rows 2 and 3. The figures for Blacks are the average of Caribbean, African and others, and the figures for South Asians are the average of Bangladeshis, Indians, and Pakistanis. Rows 12 and 13 give results for the same sample for ever use of glue, gas or solvents and shows this was highest for South Asians, intermediate for Blacks, and lowest for Whites. Rows 14 and 15 give results for the same sample for ever use of class A drugs or amphetamines and shows that for boys this was highest for South Asians, intermediate for Blacks, and lowest for Whites, while for girls this was highest for Blacks, intermediate for South Asians, and lowest for Whites. In the fifteen studies Blacks had the highest percentage of drug abuse in eight, South Asians had the highest percentage in six, while Whites had the highest percentage in one.

Table 4.15. Race differences in drug abuse (percentages)

	COUNTRY	DRUG ABUSE	BLACK	SOUTH ASIAN	WHITE	REFERENCE
1	Netherlands	Alcoholism	17.0	1.4	4.3	Selten & Sijben,1994
2	Britain: boys	Marijuana	49.3	21.5	33.5	Rodham et al., 2005
3	Britain: girls	Marijuana	32.2	5.5	28.0	Rodham et al., 2005
4	Britain: boys	Ecstasy	7.0	7.3	4.8	Rodham et al., 2005
5	Britain: girls	Ecstasy	1.9	2.3	7.5	Rodham et al., 2005
6	Britain: boys	Stimulants	9.9	7.0	6.1	Rodham et al., 2005
7	Britain: girls	Stimulants	3.1	3.3	5.0	Rodham et al., 2005
8	Britain: boys	Opiates	7.0	7.5	2.3	Rodham et al., 2005
9	Britain: girls	Opiates	2.1	3.0	1.2	Rodham et al., 2005
10	Britain: boys	Marijuana	15.8	7.4	10.9	Jayakody et al., 2006
11	Britain: girls	Marijuana	21.2	3.1	10.2	Jayakody et al., 2006
12	Britain: boys	Glue, gas	8.0	8.8	5.0	Jayakody et al., 2006
13	Britain: girls	Glue, gas	5.5	6.6	3.0	Jayakody et al., 2006
14	Britain: boys	Other	1.6	4.6	3.0	Jayakody et al., 2006
15	Britain: girls	Other	3.6	2.8	2.0	Jayakody et al., 2006

12. CHEATING

Racial differences in cheating in sport as an expression of moral weakness have been examined in English Premier League football (soccer) by Dutton and Lynn (2014). The study reported that in the 2012-2013 season, 32 percent of the Premier League footballers were Black and 67 percent were White; in the 2010-2011 season 25 percent were Black and 75 percent were White; and in 2006-2007 12 percent were Black and 88 percent were White. The study examined the players handed red cards for cheating in the three seasons. The results are given in Table 4.16, and show that in all three years Black Premier League footballers were over-represented among those given red-cards for cheating.

Table 4.16. Players receiving Red Cards in the English Premier League

SEASON	N	PERCENTAGE BLACK	PERCENTAGE GIVEN RED CARDS BLACK
2012-13	36	32	42
2010-11	40	25	37.5
2006-7	29	12	27

13. ALTRUISM

Altruism and more generally pro-social behavior being helpful to others is the antithesis of psychopathic personality. Race differences in altruism assessed by willingness to donate organs in the event of death were shown in the United States in Table 2.43 in a study reporting that 43 percent of Whites, but only 31 percent of Hispanics, and 23 percent of Blacks were willing to donate their organs after death. A study in Sweden by Sanner (1990) found that 61 percent of a random sample of adults were willing to donate their organs after death, confirming that Europeans have a high level of altruism.

Data for racial differences in pro-social behavior in Britain have been reported from teachers' assessments of 7,255 11 year olds. These gave Blacks as the average of Indians, Pakistanis and Bangladeshis, -.07*d* lower than Whites, and Chinese .22*d* higher than Whites (Lynn & Cheng, 2016).

14. CONCLUSIONS

The studies from Europe reviewed in this chapter confirm those in the United States by showing greater rates of psychopathic personality in Blacks found in the epidemiological surveys and expressed in higher rates of conduct disorders, exclusions from school, delinquency and crime, multiple sexual partners, recklessness, sexual precocity, child maltreatment, drug abuse and cheating, and lower rates of pro-social behavior. The studies from Europe also confirm those in the United States by showing lower rates of psychopathic personality in Northeast Asians expressed in lower rates of conduct disorders, exclusions from school, crime and higher rates of marriage and co-habitation. South Asians have about the same level of psychopathic personality as Europeans.

CHAPTER 5

SUB-SAHARAN AFRICA

Sub-Saharan Africans have been regarded as one of the major races in the taxonomies of classical anthropology set out in the mid-eighteenth century by of Linnaeus (1758) and Blumenbach (1776). In twentieth century anthropology they were termed *Negroids* (Coon, Garn & Birdsell, 1950). Cavalli-Sforza, Menozzi, & Piazza (1994) in their classification of humans into genetic "clusters" have confirmed the distinctive genetic characteristics of the sub-Saharan Africans who include west Africans, Nilotics in southern Sudan, Ethiopians, and Bantus, a large group present in most of Sub-Saharan Africa. The most distinctive features of Africans are their very dark skin, dark eyes, broad nose, thick averted lips, and woolly hair. Their blood groups differ from Europeans in having a lower frequency of group A, which is present in about 27 percent as compared with around 46 percent in Europeans, and a higher frequency of group B, which is present in about 34 percent as compared with around 14 percent in Europeans.

In the early and mid-twentieth century several physicians who worked in Africa described the psychopathic characteristics of sub-Saharan Africans. For instance, Carmen and Roberts (1934) wrote of a Nilotic people as "happy-go-lucky, irresponsible and living for the day." Williams (1938) wrote of the Gold Coast people as "almost invariably dishonest." Westermann (1939) described them as having "few gifts for work which aims at a distant goal and requires tenacity, independence and foresight." Barbé (1951) wrote of their "impulsivity, violent but unsustained, inconstancy, recklessness… and lack of persistent effort." And Carothers (1953) described them as "unstable, impulsive, unreliable, irresponsible, and living in the present without reflection or ambition, or regard for the rights of people outside their own circle."

1. PREVALENCE OF PSYCHOPATHIC PERSONALITY

Studies of race differences in the prevalence of psychopathic personality are summarized in Table 5.1. Row 1 gives data from an administration of the Psychopathic Deviate Scale of the Minnesota Multiphasic Personality Inventory (MMPI-2) to 200 Nigerian male and female students in Nigeria. The data are given as ds (standard deviation units) in relation to 1.0 for White American norms, and show that they obtained an average score half a standard deviation higher than White American students. Row 2 gives the results of a study of the prevalence of psychopathic personality in a national survey of a representative sample of 8,351 in Great Britain in 2000 and again shows a higher prevalence of about the same magnitude in Blacks.

Table 5.1. Black-white differences the prevalence of psychopathic personality (ds)

	COUNTRY	BLACK	WHITE	REFERENCE
1	Nigeria	1.50	1.0	Nzewi, 1998
2	Britain	1.44	1.0	Crawford et al., 2011

2. ATTENTION DEFICIT HYPERACTIVITY DISORDER

A study of race differences in attention deficit hyperactivity disorder (ADHD) in South Africa are given in Table 5.2. The results show the highest percentage of ADHD in Blacks followed by South Asians, and the lowest percentage in Whites.

Table 5.2. Race differences in attention deficit hyperactivity disorder (percentages)

BLACK	SOUTH ASIAN	WHITE	REFERENCE
7.4	4.8	3.9	Yao et al., 1988

3. CRIME

Race differences in convictions for male homicide per 100,000 population in South Africa are given in Table 5.3.

Table 5.3. Race differences in homicide per 100,000 in South Africa

	YEAR	BLACK	COLORED	ASIAN	WHITE	REFERENCE
1	1978	23.9	26.5	4.4	3.8	Lester,1989
2	1981	24.5	76.6	10.0	6.8	Lester,1989
3	1984	34.5	58.0	9.9	5.8	Lester,1989
4	1982-90 (men)	47.5	48.2	29.4	15.4	Thomson, 2004
5	1982-90 (women)	15.6	24.4	13.8	11.5	Thomson, 2004
6	2001-5	75.9	53.2	17.8	15.1	Matzopoulos et al, 2014

The rate for Coloreds (a mixed largely Black-White group) is the highest, the Black rate comes next at about six times greater than the White. The Asians consist of ethnic Indians from the Indian sub-

Continent, and is about fifty percent higher than the White rate. Rows 4 and 5 give data for men and women 16-30 year olds, the peak ages for crime explaining why the rates were higher but showing the same race differences. Row 6 gives more recent male homicide rates for five cities showing similar differences with the Black rate the highest and five times greater that the rate of Whites.

4. INTIMATE PARTNER VIOLENCE

In a study of intimate partner violence against women in Sub-Saharan Africa, Uthman, Lawoko and Moradi (2009, p.14) write,

> Intimate partner violence against women is deep-rooted in many African societies, where it is considered a prerogative of men and a purely domestic matter in the society. It is one of the greatest barriers to ending the subordination of women. Women, for fear of violence, are unable to refuse sex or negotiate safer sexual practices, thus increasing their vulnerability to HIV if their husband is unfaithful. Violence against women, especially by intimate partners, is a serious public health problem that is associated with physical, reproductive and mental health consequences. Even though most societies proscribe violence against women, the reality is that violations against women's rights are often sanctioned under the garb of cultural practices and norms, or through misinterpretation of religious tenets. Moreover, when violation takes place within the home, as it is often the case, the abuse is effectively ignored by the tacit silence and the passivity displayed by the state and the law enforcing machinery. The global dimensions of this violence are alarming as highlighted by numerous studies. A troubling aspect of intimate partner violence is its benign social and cultural acceptance of physical

chastisement of women and is the husband's right
to 'correct' an erring wife.

Studies of men's and women's attitudes toward intimate partner violence against women in seventeen countries in Sub-Saharan Africa carried out between 2003 and 2007 have been published by Uthman, Lawoko and Moradi (2009). One of the questions asked in these surveys was whether men are justified in using physical chastisement on their wives for arguing with their husbands. The percentages of men and women who gave affirmative answers are given in Table 5.4. It is a curious feature of these studies that in twelve of the countries more women than men considered intimate partner violence against women justified.

Table 5.4. Justification for intimate partner violence against women (percentages)

COUNTRY	MEN	WOMEN	COUNTRY	MEN	WOMEN
Benin	7.2	35.5	Mozambique	22.5	34.4
Burkina Faso	21.4	55.7	Namibia	24.4	17.1
Ethiopia	27.6	51.0	Nigeria	24.4	41.6
Ghana	18.2	32.0	Rwanda	3.0	6.5
Kenya	40.9	45.8	Swaziland	22.7	17.5
Lesotho	38.7	37.2	Tanzania	24.1	42.7
Liberia	23.0	40.9	Uganda	36.6	40.0
Madagascar	3.8	2.9	Zimbabwe	21.4	26.7
Malawi	8.1	12.0			

Studies of rates of intimate partner violence experienced by married women in the last year in Africa are given in Table 5.5. These percentages are much higher than those of around 3 to 12 percent for White women in the United States and Britain.

Table 5.5. Rates of intimate partner violence (percentages)

	COUNTRY	YEAR	IPV	REFERENCE
1	Egypt	1995	34.4	Kishor & Johnson, 2004
2	Ethiopia	2000	49.0	WHO, 2002
3	Nigeria	2008	18.7	Linos et al., 2013
4	Uganda	2000	30.0	Koenig et al., 2003
5	Zambia	2001	48.4	Kishor & Johnson, 2004

A study of race differences in rates of sexual violence experienced by women during the preceding ten years in South Africa reported by their male partners is summarized in Table 5.6. The data are for 1998 and are expressed as odds ratios with the rate for Blacks set at 1.00, and show the rate for Blacks approximately eight times greater than that for Whites, and the rate for Blacks slightly greater than the rate for Coloreds and Indians.

Table 5.6. Rates of intimate partner violence in South Africa (odds ratios)

BLACKS	COLOREDS & INDIANS	WHITES	REFERENCE
1.00	0.93	0.12	Abrahams et al., 2004

5. SEXUAL PRECOCITY

Early sexual experience as a feature of psychopathic personality has been reviewed by Kaaya et al. (2002, p. 30) who concluded that "for school students of both genders in Sub-Saharan Africa, there is an early age of the onset of sexual behavior and more than one lifetime sexual partner." Studies providing data on age of first intercourse in a number of African countries are summarized in Table 5.7.

Table 5.7. Age of first intercourse

	COUNTRY	N	SEX	AGE	REFERENCE
1	Burundi	3970	F	19.5	Blanc & Rutenberg, 1993
2	Ghana	4488	F	16.5	Blanc & Rutenberg, 1993
3	Kenya	7150	F	16.5	Blanc & Rutenberg, 1993
4	Malawi	2459	M	10.5	Mkandawire et al, 2013
5	Nigeria	208	MF	11.4	Lawrence et al., 1995
6	Nigeria	1491	M	14.6	Slap et al., 2003
7	Nigeria	1244	F	15.2	Slap et al., 2003
8	Tanzania	518	M	15.5	Munguti et al., 1997
9	Tanzania	599	F	15.5	Munguti et al., 1997
10	Tanzania	482	M	16.0	Matasha et al., 1998
11	Tanzania	308	F	16.0	Matasha et al., 1998
12	Uganda	4730	F	15.5	Blanc & Rutenberg, 1993
13	Africa: median	-	M	15.0	
14	Africa: median	-	F	15.5	

	COUNTRY	N	SEX	AGE	REFERENCE
15	USA- Blacks	-	M	14.1	Table 2.35
16	USA- Blacks	-	F	16.4	Table 2.35
17	USA- Whites	-	M	16.4	Table 2.35
18	USA- Whites	-	F	17.0	Table 2.35

Rows 1 through 12 give results for age of first intercourse for seven countries in Sub-Saharan Africa. Rows 13 and 14 give the medians for these results. Rows 15 and 16 give the means for Blacks in the United States taken from Table 2.35 showing broadly similar ages of first intercourse. Rows 17 and 18 gives the means for Whites in the United States showing later ages of first intercourse. Studies providing data on age of first intercourse as percentages are summarized in Table 5.8. Rows 1 and 2 give percentages for males and females who had engaged in intercourse by the ages of 13/14, 14/15, and 16/17, obtained in a 1995 survey in Botswana. Rows 3 through 6 give similar figures for males and females for Swaziland and Tanzania obtained in surveys carried out in 1988 and 1991. Figures for Blacks and Whites in the United States are given in rows seven through ten, showing much higher percentages at these young ages for Blacks than for Whites.

Table 5.8. Age of first intercourse (percentages)

	COUNTRY	N	SEX	AGE 13/14	AGE 15/16	AGE 17/18	REFERENCE
1	Botswana	1174	M	4.7	40.8	70.0	Meekers & Ahmed, 2000
2	Botswana	1203	F	1.4	14.7	65.0	Meekers & Ahmed, 2000

	COUNTRY	N	SEX	AGE 13/14	AGE 15/16	AGE 17/18	REFERENCE
3	Swaziland	40	M	-	57.5	85.0	McLean, 1995
4	Swaziland	38	F	-	64.4	0.0	McLean, 1995
5	Tanzania	529	M	35.8	-	-	Ndeki et al., 1994
6	Tanzania	590	F	15.1	-	-	Ndeki et al., 1994
7	US: blacks	-	M	49	-	-	Coker et al., 1994
8	US: blacks	-	F	12	-	-	Coker et al., 1994
9	US: whites	-	M	18	-	-	Coker et al., 1994
10	US: whites	-	F	3	-	-	Coker et al., 1994

Further data on precocious sexuality in Sub-Saharan Africa, consisting of the percentages of school students who had had sexual intercourse by age of fifteen years, are given in Table 5.9 together with comparable data from the UK.

Table 5.9. Race differences in sexual intercourse by age of fifteen years (percentages)

	COUNTRY	M	F	REFERENCE
1	Kenya	48	17	Kaaya et al., 2002
2	Namibia	56	17	Kaaya et al., 2002
3	Tanzania	63	24	Kaaya et al., 2002
4	UK: Blacks	26	10	Wellings et al., 1994
5	UK: Whites	19	9	Wellings et al., 1994

6. MULTIPLE SEXUAL PARTNERS

Multiple sexual partners are common in Sub-Saharan Africa both as polygamous marriages, and informally where "the decline in polygamous unions in the urban areas of Sub-Saharan Africa has been accompanied by the growth of various forms of multiple and/or serial informal marriages which involve rather irregular "girlfriends" and somewhat regular "outside wives." (Hayase & Liaw, 1997, p. 293). Polygyny in Sub-Saharan Africa varies between 11.6 percent of married women in Burundi to 52.3 percent of married women in Togo (Speizer & Yates, 1998). In Kenya, 29.5 percent of married women were in polygynous unions in 1977, falling to 19.5 percent in 1993 (Ezeh, 1997). A study in South Africa attributes the high rate of HIV infection to "a traditional custom whereby people have more than one lover." (Fourie, 2004, p.254)

Studies of multiple sexual partners for adults in Africa and comparable data for Britain are shown in Table 5.10.

Table 5.10. Multiple sexual partners (percentages)

	COUNTRY	SEX	PAST YEAR 5+	PAST 6 MONTHS 2+	REFERENCE
1	Tanzania	M	9	-	Munguti et al., 1997
2	Tanzania	F	1	-	Munguti et al., 1997
3	Britain	M	2	-	Johnson et al. 1994
4	Britain	F	0	-	Johnson et al. 1994
5	Zaire	M	-	23	Bertrand et al., 1991
6	Zaire	F	-	1	Bertrand et al., 1991

Rows one and two give percentages of those who had engaged in sexual intercourse with five plus partners in the past year for Tanzania, and rows three and four give lower figures for Whites in Britain. Rows five and six give the percentages in Zaire (now the Democratic Republic of Congo) of those engaged with two or more partners in the last six months.

Table 5.11 gives data for multiple sexual partners for adolescents aged 13-18 in the last year in Botswana, reported by Meekers and Ahmed (2000), and comparable data for 15-19 year olds in the USA, reported by Moore et al. (1998), and show higher percentages in Blacks than in Whites.

Table 5.11. Multiple sexual partners in the last year (percentages)

	COUNTRY	SEX	N PARTNERS 0	N PARTNERS 1	N PARTNERS 2+
1	Botswana	M	15.2	21.8	63.0
2	Botswana	F	24.1	19.3	56.7
3	USA: blacks	M	-	-	63
4	USA: blacks	F	-	-	33
5	USA: whites	M	-	-	39
6	USA: whites	F	-	-	26

7. LONG-TERM MONOGAMOUS RELATIONSHIPS

Marriage is an index of the ability to form long-term monogamous relationships that is lacking in psychopathic personality. Data for greater percentages never married among 35-50 year old men in a 2002-4 survey in two sub-Saharan African and two European countries are given by Ambugo (2014), and are shown in Table 5.12. Table 5.12. Percentages never married

COUNTRY	N	PERCENT
Ghana	3922	24.8
Kenya	4331	26.9
Norway	943	21.0
UK	1195	20.3

A further expression of the weakness of the ability to form long-term monogamous relationships in Sub-Saharan Africa is the higher divorce rates than in Western countries. For example, divorce rates over the first 5 years of marriage are 39 percent in the !Kung of the Kalahari Desert (Howell, 1979), and 37 percent in the Hadza (Blurton Jones, Marlowe, Hawkes & O'Connell, 2000). The equivalent current divorce rates for the first 5 years from national probability samples in the United States are 17 percent (Manning & Cohen, 2012).

8. RECKLESSNESS: ON-USE OF CONTRACEPTION

There has been a high prevalence of HIV and AIDS in Sub-Saharan Africa from the 1980s into the twenty-first century. In 1999, 70 percent of the world population with this infection and condition were in sub-Saharan Africa (Messersmith et al., 2000). In the high risk urban areas of Eastern and Southern Africa, and in Nigeria, the rate of infection in the 1990s was around 20 to 30 percent (UN AIDS, 2000). It is reckless not to use condoms in Sub-Saharan Africa where there is a high rate of HIV infection, yet there is a high rate of unprotected sexual intercourse in Sub-Saharan Africa. A review of a number of studies by Kaaya et al. (2002, p. 30) concluded that, "for school students of both genders in Sub-Saharan Africa, a large proportion of the sexually active report unprotected sexual intercourse." A number of studies showing this are summarized in Table 5.13.

Table 5.13. Use of condoms in sub-Saharan Africa (percentages)

	COUNTRY	SEX	USE CONDOMS	REFERENCES
1	Cameroun	M	18.7	Raji & Adegboye,1993
2	Ghana	F	18.9	Klomegah, 1999
3	Malawi	M	40.0	Mkandawire et al, 2013
4	Nigeria	M	33.1	Messersmith et al., 2000
5	South Africa	F	42.5	Kapiga et al., 1993
6	Tanzania	MF	29.9	Mnyika et al., 1995
7	Uganda	M	9.9	Lutalo et al, 2000
8	Uganda	F	4.4	Lutalo et al, 2000
9	Zaire	M	24.0	Bertrand et al.,1991
10	Zaire	F	11.8	Bertrand et al.,1991

Row 1 gives 18.7 percent of men in Cameroon who were clients of prostitutes who said they used condoms, the remaining 81.3 percent saying they did not use condoms because these reduced sexual pleasure. Row 2 gives 18.9 percent of women aged 15-49 in a 1993 study in Ghana who used any form of contraception. Row 3 gives 40 percent of men Malawi who said they had used a condom at first intercourse. Row 4 shows 33.1 percent of men in Nigeria who had ever used a condom for STD protection. Row 5 gives 42.5 percent of Black women in South Africa who said they used condoms, the remaining 57.5 percent saying they did not use condoms because these reduced their men's sexual pleasure. Row 6 gives 29.9 percent of men and women in Tanzania in a 1993 study who had used condoms regularly in a sample of 748 who knew what condoms are for. A further 333 individuals in this study did not know what condoms are for. Rows 7 and 8 give 9.9 percent of men and 4.4 percent of women in Uganda who used condoms in a 1995 study in which condoms were given out and the sample were given counseling on the possibility of contacting HIV and other STDs through unprotected sexual intercourse. Rows 9 and 10 give the use of condoms in Zaire by those with extra-marital partners as 24 percent for men and 11.8 percent for women.

9. SEXUALLY TRANSMITTED DISEASES

It was noted in Chapter 2 that in the United States there is an association between race differences in psychopathic personality and rates of sexually transmitted diseases (STDs). The high rates of unprotected sexual intercourse in Sub-Saharan Africa would be expected to lead to high rates of HIV infection, and to race differences in the rates of other sexually transmitted diseases (STDs). This was been found in a 2008 study in South Africa whose results are given in Table 5.14, showing that Blacks have approximately eight times the rate of HIV infection as Coloreds, and approximately 45 times the rate of HIV infection as Asians and Whites.

Table 5.14. Race differences in HIV infection in South Africa (percentages)

ASIAN	BLACK	COLORED	WHITE	REFERENCE
0.3	13.6	1.7	0.3	Human Sciences Res., 2009

10. WORK COMMITMENT

Studies of racial and ethnic differences in work motivation and commitment in the United States were summarized in Table 2.27, and showed that this was weaker in Blacks and Hispanics in Whites and Asians. Similar differences have been found in Africa in comparisons between Blacks and ethnic Indians originally from South Asia. In the nineteenth century the European colonists needed laborers for agricultural and manual work of various kinds. Initially they tried to employ Blacks but they found that these would not work reliably and brought over Indians who were more satisfactory. Sir Harry Johnston, a British colonial administrator, explained the problem:

> These semi-tropical plantations brought about a fresh want—that of patient, cheap, agricultural laborers. Unhappily, the black man, though so strong in body and so unaspiring in ideals, has as a rule a strong aversion to continuous agricultural labor. His own needs are amply supplied by a few weeks' tillage scattered throughout the year; and even this is generally performed by the women of the tribe, the men being free to fight, hunt, fish, tend cattle, and loaf. Therefore the black men of Natal, though they made useful domestic servants and police, were of but little use in the plantations. (Johnston, 1930, p. 271).

Some two decades later Leonard Thompson (1952, p. 5) wrote, "that a labour problem should have existed in Natal may at first sight seem inconceivable, but though there were a hundred thousand natives in Natal there were not enough labourers." More recently University of Toronto historian Rick Halpern (2004, p. 25) has written in a similar vein: "the work routine at harvest time—necessitating around the clock operations—ran counter to indigenous conventions." The same problem surfaced at the end of the nineteenth century in East Africa when the British in Kenya and the Germans in Tanganyika needed laborers to build the railroads. They found that it was not possible to get Blacks to do the work. The Blacks "were content to live with little effort at subsistence level and did not want to work for whites." (K.I., 1960, p. 342) The Germans also encountered this problem when attempting to use Blacks to build the railroad in Tanganyika and the Portuguese had a similar experience in their colony of Mozambique. They both failed and adopted the solution of bringing in Indians for this work. The British psychologist Philip Vernon has summed up these experiences, writing that, in Africa, "work is generally leisurely and periodic, depending on the climate, the rhythms of nature and local custom; regularity or an accurate sense of time are unimportant so that, to whites, the African seems indolent." (Vernon 1969, p. 177)

11. CHILD MALTREATMENT

The results of a study of the prevalence of self-reported childhood physical abuse and neglect experienced by Kenyan, Zambian, and Dutch university students are summarized in Table 5.15. Row 1 gives the numbers, row 2 gives the percentages reporting childhood physical abuse, showing much higher percentages in the Kenyans and Zambians than in the Dutch. Row 3 gives the percentages reporting childhood neglect and again showing higher percentages in the Kenyans and Zambians than in the Dutch.

Table 5.15. Child maltreatment (percentages)

	MALTREATMENT	KENYA	ZAMBIA	NETHERLANDS	REFERENCE
1	Numbers	375	182	305	Mbagaya1 et al, 2013
2	Physical abuse	42	40	3	Mbagaya1 et al, 2013
3	Neglect	59	54	42	Mbagaya1 et al, 2013

12. CONCLUSIONS

The studies in Sub-Saharan Africa summarized in this chapter show a high level of psychopathic personality expressed in high rates of attention deficit hyperactivity disorder, homicide, sexual precocity, multiple sexual partners, inability to form long-term monogamous relationships, recklessness in the non-use of contraception, HIV infection and child maltreatment, and low work commitment. Europeans and South Asians show lower levels of psychopathic personality consistent with results in the United States, Canada and Europe.

1. Prevalence

2. Conduct disorder, ADHD and Psychopathic Personality

3. Crime

4. Moral Values

5. Sexual Precocity

6. HIV Infection

7. Intimate Partner Violence

8. Multiple Sexual Partners and Extramarital Sex

9. Drug and Substance Abuse

10. Altruism

11. Conclusions

CHAPTER 6

NORTHEAST ASIA

The Northeast Asians are the indigenous peoples of China, Japan, Korea, and Mongolia. In classical anthropology they are described as Mongoloids and have been recognized as one of the major races from the first taxonomies of Linnaeus (1758) and Blumenbach (1776), and as one of the seven major races in the classification proposed by Coon, Garn, and Birdsell (1950). Their identity as a genetic "cluster" has been confirmed by Cavalli-Sforza, Menozzi and Piazza (1994) in their classification based on a number of genetic markers taken from samples of Samoyeds, Mongols, Tibetans, Koreans, and Japanese. The most distinctive features of Northeast Asians are their straight black hair, flat nose, yellowish skin color and the epicanthic eye-fold that gives their eyes a narrow appearance.

1. PREVALENCE

Epidemiological studies of lifetime prevalence rates of psychopathic personality in three Northeast Asian and ten European countries and the United States are given in Table 6.1, and show a lower average rate of 1.5 in the Northeast Asian countries than 3.86 in European populations.

Table 6.1. Lifetime prevalence rates of psychopathic personality (percentages)

COUNTRY	RATE	REFERENCE
Canada	3.7	Bland et al., 1988
Canada	2.9	Weissman, 1993
Denmark	0.5	Weissman, 1993
Iceland	4.0	Weissman, 1993
New Zealand	2.4	Wells et al., 1989
New Zealand	3.1	Weissman, 1993
Hong Kong	1.4	Chen et al., 1993
Norway	9.4	Weissman, 1993
Sweden	5.6	Weissman, 1993
South Korea	2.2	Lee et al., 1990
Taiwan	0.8	Weissman, 1993
USA	2.7	Robins et al., 1991
USA	4.3	Samuels et al. 2002

2. CONDUCT DISORDER, ADHD AND PSYCHOPATHIC PERSONALITY

Studies of Northeast Asian-American comparisons on the prevalence of psychopathic personality, and the associated conditions of conduct disorder and attention deficit hyperactivity disorder (ADHD), are summarized in Table 6.2. Row 1 shows the rate of conduct disorder lower in Hong Kong than in the United States. The data are given as ds (standard deviation units) in relation to zero for the United States. Row 2 shows the rate of attention deficit hyperactivity disorder lower in Hong Kong than in the United States. Row 3 shows the percentages in the populations with anti-social personality disorder lower in Taiwan than in the United States. Row 4 shows the percentage of the population with psychopathic personality is lower in Japan than among Whites in the United States according to the data from the Psychopathic Deviate Scale of the Minnesota Multiphasic Personality Inventory (MMPI) given in the Japanese and American standardization manuals of the MMPI-2. The Japanese norms are based on a nationally representative sample of 1,022.

Table 6.2. Northeast Asian-American comparisons on measures of psychopathic personality

	COUNTRY	CONDITION	NORTHEAST ASIA	UNITED STATES	REFERENCE
1	Hong Kong	Conduct disorder: d	-.49	.00	Luk & Leung, 1989
2	Hong Kong	ADHD: d	-1.0	.00	Luk & Leung, 1989
3	Taiwan	ASPD:%	0.18	2.53	Compton, et al., 1991
4	Japan	Psychopathy: d	-.36	.00	Japanese MMP1, 1993

3. CRIME

Crime rates for homicide and rape and per 100,000 population for five Northeast Asian countries for the years 1993-1996 have been given by Rushton and Templer (2009). Their results are given in Table 6.3 together with the medians of the rates averaged for 36 European countries. It will be seen that for both crimes the rates in the Northeast Asian countries are considerably lower than in Europe.

Table 6.3. Rates of homicide and rape per 100,000 population for 5 Northeast Asian and countries and Europe

COUNTRY	HOMICIDE	RAPE
China	1	3
Hong Kong	1	2
Japan	1	1
Singapore	1	3
South Korea	1	3
Europe	5	6

4. MORAL VALUES

The Defining Issues Test has been developed as a measure of moral values by Rest (1979, 1986). This consists of a number of stories in which the leading actor is confronted with a moral dilemma. The problem is to discern the moral principle involved, and choose the course of action consistent with it. This test has been given to high school students in the age groups 14-18 in South Korea and Hong Kong. The results are given in Table 6.4 expressed as *d* scores for

154

these national samples in relation to American scores of zero. In both studies South Korea and Hong Kong students obtained higher scores showing stronger moral values than those in the United States.

Table 6.4. Race differences in moral values assessed by the Defining Issues Test (ds)

	COUNTRY	NORTHEAST ASIAN	EUROPEAN	REFERENCE
1	South Korea	1.71	0.00	Park & Johnson, 1984
2	Hong Kong	1.31	0.00	Rest, 1986

5. SEXUAL PRECOCITY

Studies providing data on sexual precocity as an expression of psychopathic personality in young adolescents are summarized in Table 6.5.

Table 6.5. Experience of sexual intercourse among university students (percentages)

	COUNTRY	SEX	NORTHEAST ASIAN	EUROPEAN	REFERENCE
1	Hong Kong	M	3.5	-	Fan et al., 1995
2	Hong Kong	F	1.4	-	Fan et al., 1995
3	China	M	6.3	-	Fan et al., 1995
4	China	F	2.9	-	Fan et al., 1995
5	Britain	M	-	50.0	Wellings et al., 1994

	COUNTRY	SEX	NORTHEAST ASIAN	EUROPEAN	REFERENCE
6	Britain	F	-	50.0	Wellings et al., 1994
7	Australia	M	40	67	Rosenthal et al., 1990
8	Australia	F	22	68	Rosenthal et al., 1990

Rows 1 and 2 give the percentages of male and female 18-21-year-old university students who had engaged in sexual intercourse in a study of 4,688 students in Hong Kong. Rows 3 and 4 give the percentages of male and female university students who were sexually active in a study of 481 students in Shanghai. Rows 5 and 6 give much higher percentages of male and female university students who had copulated in Britain in a study carried out in the same year. Rows 7 and 8 give lower percentages of male and female Chinese university students than Europeans who had were sexually active in Australia.

For both men and women, significantly higher percentages of Anglo-Australian than of Chinese-Australian students had experience of sexual intercourse. Resulted are summarized in Table 6.6.

Table 6.6. College students experience of sexual intercourse (percentages)

	ANGLO	AUSTRALIAN	REFERENCE
Men	67	40	
Women	68	22	

6. HIV INFECTION

High levels of psychopathic personality would be expected to lead to high rates of HIV infection and it would therefore be expected that these would be low in Northeast Asia. This is shown in Table 6.7 which gives rates of HIV infection for 2009 for adults aged 16-45 years in ten world regions, showing that these were lowest in Northeast Asia (UN AIDS, 2009). This table also shows the highest rates of HIV infection in Sub-Saharan Africa and the Caribbean, providing further evidence for high levels of psychopathic personality in these populations.

Table 6.7. HIV infection rates in ten world regions (percentages)

REGIONS	HIV %	REGIONS	HIV %
Sub-Saharan Africa	5.0	South/Southeast Asia	0.3
Caribbean	1.0	Oceana	0.3
East Europe/Central Asia	0.8	West/Central Europe	0.2
South/Central America	0.5	North Africa/Middle East	0.2
North America/Mexico	0.5	Northeast Asia	0.1

7. INTIMATE PARTNER VIOLENCE

Studies of last year prevalence of intimate partner violence against women in Japan and the United States are summarized in Table 6.8. Row 1 gives results from a 1998 study of 4,500 women in Japan showing 2.7 percent had experienced intimate partner violence during the last year. Row 2 gives the average of five studies of intimate partner violence by White men against women in the United States presented in Table 2.24, showing that this is substantially higher than in Japan.

Table 6.8. Intimate partner violence (IPV) against women (percentages)

	COUNTRY	IPV	REFERENCE
1	Japan	2.7	Sourifu, 1999
2	United States	4.4	Table 2.24

8. MULTIPLE SEXUAL PARTNERS AND EXTRAMARITAL SEX

The desire for multiple sexual partners is one the classical defining features of psychopathic personality (see Chapter 2, section 10). It has been shown that this desire is weaker in Northeast Asians than in Europeans in a study that asked college students the number of sexual partners they desired in the next month and whether they desired more than one sexual partner (Schmitt, 2003). The results of this study are given in Table 6.9.

Table 6.9. Desire for multiple sexual partners

REGIONS	MEN N DESIRED	WOMEN N DESIRED	MEN 1+ %	WOMEN 1+ %
North America	1.66	0.63	23.1	2.9
West Europe	1.82	0.93	27.6	2.9
East Europe	2.43	1.01	31.7	7.1
South Europe	2.01	0.90	31.0	6.0
South Asia	1.98	0.88	32.4	6.4
Northeast Asia	1.25	0.35	17.9	2.6

It will be seen that Northeast Asians (in Hong Kong, Japan, South Korea and Taiwan) scored lower than the four samples of Europeans, and South Asians scored about the same as Europeans on these measures. These results are corroborated by a study showing that infidelity during the last year is relatively rare at 8 percent of men and 1 percent of women in Hong Kong in comparison with 38 percent of men and 19 percent of women in Guinea Bissau (Africa) (Careal, Cleland, Deheneffe, Ferry & Ingham, 1995).

9. DRUG AND SUBSTANCE ABUSE

Studies of drug and substance abuse in Northeast Asia and the United States are summarized in Table 6.10. Row 1 shows the prevalence of alcohol abuse substantially lower Taiwan than in the United States. Row 2 shows the prevalence of drug abuse substantially lower Taiwan than in the United States. Row 3 shows the percentage of those who have ever used stimulants, cocaine or opioids in surveys during 1981 to 1986 are substantially lower in Japan and than in the United States.

Table 6.10. Drug and substance abuse in Northeast Asia and the United States (percentages)

	COUNTRY	MEASURE	NORTHEAST ASIA	UNITED STATES	REFERENCE
1	Taiwan	Alcohol abuse	7.18	13.43	Compton, et al., 1991
2	Taiwan	Drug abuse	0.16	6.12	Compton, et al., 1991
3	Japan	Stimulants	0.9	7.1	Price at al., 1995

10. ALTRUISM

Willingness to donate organs in the event of death was adopted to assess race differences in altruism in the United States in Chapter 2 and the results of a number of studies showing higher rates in Asians and Europeans were summarized in Table 2.43. The American results obtained from a telephone survey found that 43 percent of Whites, but only 31 percent of Hispanics, and 23 percent of Blacks were willing to donate their organs after death. A similar study of 284 adults also obtained from a telephone survey carried out by Yeung, Kong, and Lee (2000) in Hong Kong in 1996 found that 60.3 percent were willing to donate their organs after death providing further evidence for a high level of altruism in Northeast Asians.

11. CONCLUSIONS

The studies of Northeast Asians summarized in this chapter show a low level of psychopathic personality expressed in epidemiological studies, low rates of conduct disorder, attention deficit hyperactivity disorder, homicide, sexual precocity, moral values, HIV infection, the desire for multiple sexual partners and drug and substance abuse, and a high level of altruism. These results confirm the studies showing a low level of psychopathic personality in Northeast Asians in the United States and Europe.

CHAPTER 7

SOUTH ASIA AND NORTH AFRICA

The South Asians and North Africans are the indigenous peoples of southern Asia from Bangladesh in the east through India, Pakistan, Iraq, Iran, the Gulf states, the Near East and Turkey, and of North Africa, north of the Sahara desert. They are closely related to the Europeans and in some of the taxonomies of classical anthropology, such as that of Coon, Garn and Birdsell (1950), the two peoples have been regarded as a single race designated the Caucasoids. But Cavalli-Sforza, Menozzi, & Piazza (1994), in their genetic analysis of human differences, have shown that the South Asians and North Africans form a distinctive genetic "cluster" that differentiates them from the Europeans. They are therefore treated here as a separate race.

1. GENERAL POPULATION STUDIES

A study of the prevalence of psychopathic personality in Iran in a national survey of a representative sample of 375 young males aged 16-19 years reported a rate of 6.9 percent (Aghaei & Golparvar, 2014). This can probably be regarded as about the same as the rate for young males in the United Sates and Europe although there are no data strictly comparable to the Iranian result.

Numerous studies of the prevalence of psychopathic personality in the South Asians and North Africans in Europe compared with Europeans are reviewed in Chapter 4. These show that

South Asians have somewhat higher rates of psychopathic personality than Europeans assessed in the epidemiological study in Britain given in Table 4.1, and higher rates of conduct disorders (Tables 4.2 and 4.3) and child maltreatment in three studies in the Netherlands (Table 4.16). However, the rates for crime given in Tables 4.4, 4.5, 4.6, and 4.7 are about the same as those for Europeans. South Asians have consistently higher rates of monogamous relationships assessed by rates of marriage and cohabitation (Tables 4.8 and 4.9) and lower rates of multiple sexual partners (Tables 4.10 and 4.11). South Asians have lower rates of recklessness assessed by the non-use of contraception by those who do not wish to have children (Table 4.13), lower rates of sexual precocity (Table 4.14), lower rates of intimate partner violence in Britain but a higher rate among North Africans in Spain (Table 4.15), and a lower rate of alcoholic abuse in the Netherlands (Table 4.17). A study of second generation Moroccan immigrants in Belgium reported a non-significant difference of eight percent with conduct disorders compared with six percent among Europeans (Fossian et al. (2002). A study of Asian and Black 11 year olds in Mauritius found Blacks scored higher on both aggressive and non-aggressive antisocial personality (Raine et al. 1997).

2. INTIMATE PARTNER VIOLENCE

Studies of the prevalence of last year intimate partner violence against women in South Asia are summarized in Table 7.1. The percentages are all much higher than those of around 3 to 12 percent for White women in the United States and Britain with the exception of the lower rate of 8.1 percent for Jews in Israel, about half of whom are of European origin. A lifetime prevalence of intimate partner violence of 22 percent experienced by Thai women in Sweden is given by Fernbrant et al. (2014).

Table 7.1. Experience of intimate partner violence (IPV) against women (percentages)

	COUNTRY	YEAR	IPV	REFERENCE
1	Cambodia	2000	17.5	Kishor & Johnson, 2004
2	Israel	1988	11.0	Pagovich, 2004
3	Israel: Jews	2005	8.1	Boy & Kulczycki, 2008
4	Israel: Bedouin	2005	48.3	Boy & Kulczycki, 2008
5	India	1999	18.9	Kishor & Johnson, 2004
6	Jordan	2003	42.5	Khawaja & Barazi, 2005
7	Lebanon	2005	22.0	Boy & Kulczycki, 2008
8	Syria	2005	23.1	Boy & Kulczycki, 2008
9	Thailand	2000	34.0	WHO, 2002
10	Turkey	2005	50.9	Boy & Kulczycki, 2008
11	Vietnam	2006	31.0	Vung et al., 2008
12	Vietnam	2012	27.3	Fisher et al., 2013

3. ATTENTION DEFICIT HYPERACTIVITY DISORDER

The prevalence of ADHD in 321 Indian children aged 5-10 years has been reported by Gada (1987) as 8.1 percent with a 7.6:1 ratio of boys to girls. Prevalence rates for Whites in two studies in the United States are given in Table 2.11 as 15.2 and 1.0 and are therefore inconsistent but taken together suggest that the prevalence among Indian children is about the same as that in American Whites.

4. SOUTH ASIANS IN SOUTH AFRICA

Studies giving data for rates of the prevalence of psychopathic personality in South Asians and Europeans in South Africa are reviewed in Chapter 5. These showed that South Asians had higher rates of attention deficit hyperactivity disorder at 4.2 percent compared with 3.9 percent for Europeans given in Table 5.2, and consistently higher rates of homicide than for Europeans in six studies given in Table 5.3. But South Asians had the same rates of HIV infection as Europeans given in Table 5.14.

5. CONCLUSIONS

It is difficult to draw a conclusion regarding the prevalence of psychopathic personality in the South Asians and North Africans compared with Europeans. It is considered that the best provisional conclusion is that the rate is about the same.

CHAPTER 8

THE CARIBBEAN

The Caribbean is largely populated by Blacks and mulattoes who have mixed Black and White ancestry. The percentages of Blacks and mulattoes in the major Caribbean islands are given in Table 8.1. The percentage of White ancestry in the mulattoes is given as around seven percent in Jamaica (Parra, Marcini & Akey, 1998), and around 25 percent in Trinidad (Saha & Samuel, 1987).

1. CRIME

Crime rates for homicide and rape per 100,000 population for 10 major Caribbean islands for the years 1993-1996 have been given by Rushton and Templer (2009). Their results are given in Table 8.1 together with the medians of the rates averaged for 36 European countries. It will be seen that for both crimes the rates in the Caribbean islands are higher than in Europe, with the exception of homicide in Antigua, and in most cases much higher.

Table 8.1. Percentage blacks and crime rates per 100,000 population for Europe and 10 Caribbean islands

	BLACKS %	HOMICIDE	RAPE
Europe	-	5	6
Antigua	91	5	85

	BLACKS %	HOMICIDE	RAPE
Bahamas	80	14	48
Barbados	80	6	26
Bermuda	61	14	48
Dominica	87	8	28
Dominican Rep	84	14	14
Grenada	84	8	43
Jamaica	76	31	31
Saint Kitts	-	9	40
Trinidad	40	12	11

More recent data for the rates for homicide were given as 16 per 100,000 population for the 24 Caribbean countries for the years 2003-2008 compared with 5 per 100,000 population in 42 European countries by Harrendorf, Heiskanen and Malby (2010).

Race differences in rates of crime in Cuba are summarized in Table 8.2.

Table 8.2. Race differences in crime in Cuba

	CRIME	BLACK	MULATTO	WHITE	REFERENCE
1	Population percent	12	22	66	Phillips, 1996
2	In prison percent	80	-	10	Fontova, 2005
3	Crime odds ratios	-	1.2	1.0	de Salas del Valle, 2009

Row 1 gives the percentages in the population in 1996. Row 2 shows Blacks greatly over-represented in prison. Row 3 gives data for Blacks and Mulattoes combined showing these have a higher crime rate than Whites.

2. DELAY OF GRATIFICATION

Three studies have investigated race differences in delay of gratification in the Caribbean. The first to report differences between Blacks and Whites in the delay of gratification was carried out by Mischel (1961a,b) in Trinidad. He offered Black and White children the choice between being given a small candy bar now or a larger one in a week's time. He found that Black children were much more likely to ask for the small candy bar now. This difference was confirmed in a study by Green (1964, 1972) who repeated the experiment on Black and White children in Trinidad and found the same result. A third study was carried out by Vernon (1969), who gave a delay of gratification test to Black 11-year-old boys in Jamaica and found that they had a stronger preference for instant gratification than Whites.

3. MORAL VALUES

The Defining Issues Test was developed by Rest (1979, 1986). This consists of a number of stories in which the leading actor is confronted with a moral dilemma. The problem is to discern the moral principle involved, and choose the course of action consistent with it. This test has been given to high school students in several countries with different racial populations, and the results for Trinidad and Jamaica expressed as d scores for these national samples in relation to American scores of zero are shown in Table 8.3. It will be noted that high school students in the predominantly Black populations of Jamaica and Trinidad obtain lower scores than Americans.

Table 8.3. Moral values assessed by the Defining Issues Test (ds)

	LOCATION	BLACK	WHITE	REFERENCE
1	Trinidad	-0.45	0.00	Rest, 1986
2	Jamaica	-0.51	0.00	Gielen et al., 1989

4. MONOGAMY AND MARRIAGE

There are low rates of stable monogamous unions among Blacks in the Caribbean islands. In a review of the literature, Ram & Ebanks (1973, p. 143) wrote that, "In the Caribbean in general . . . there is a substantial amount of movement from one sex partner to another and also a very high percentage of reproduction outside marriage."

Racial differences in stable monogamous unions in Jamaica are shown in the percentages of the total population children who were married given in the 1943 census. These were highest in Europeans (42.5 percent) followed by East Indians (27.0 percent), Mulattoes (20.7 percent) and lowest in Blacks (15.5 percent).

5. SEXUAL PRECOCITY

Studies giving data on age of first sexual intercourse in Caribbean countries are summarized in Table 8.4. Rows 1, 2 and 3 give the average ages in Costa Rica and Jamaica. Rows 4 and 5 give the average ages in further studies of the age of first sexual intercourse in Jamaica. Row 6 gives 14 years as the median age at which girls had first sexual intercourse in the U.S. Virgin Islands obtained in surveys over the years 1972-86. A point of interest is the early age of boys' first sexual intercourse at 13.4 and 14.3 years in the two studies in Jamaica. Rows 7 and 8 give the average ages of the age of first sexual

intercourse of 13 for boys and 14 for girls of Blacks in Haiti. Rows 9
and 10 give the average ages of the age of first sexual intercourse of
Blacks in the United States, showing that these are about the same as
those in the Caribbean. Rows 11 and 12 give the average ages of the
age of first sexual intercourse of Whites in the United States, showing
that these are later than those of Blacks in the Caribbean and in the
United States.

Table 8.4. Age of first sexual intercourse

	YEARS	COUNTRY	SEX	AGE	REFERENCE
1	1980s	Costa Rica	F	16.6	Morris,1988
2	1980s	Jamaica	M	14.3	Morris,1988
3	1980s	Jamaica	F	16.9	Morris,1988
4	1997	Jamaica	M	13.4	Friedman et al., 1997
5	1997	Jamaica	F	15.9	Friedman et al., 1997
6	1972-86	Virgin Islands	F	14	Donoghue, 1992
7	2012	Haiti	M	13.0	Carver et al., 2014
8	2012	Haiti	F	14.5	Carver et al, 2014
9	1990	USA: Blacks	M	14.3	Day, 1992
10	1990	USA: Blacks	F	16.8	Day, 1992
11	1990	USA: Whites	M	16.3	Day, 1992
12	1990	USA: Whites	F	17.4	Day, 1992

6. RECKLESSNESS: NON-USE OF CONTRACEPTION

Studies providing data on the low percentages of boys and girls that used contraception during first sexual intercourse in Jamaica in the 1980s are summarized in Table 8.5.

Table 8.5. Use of contraception in Jamaica (percentages)

SEX	AGE	PERCENT USE	REFERENCE
Boys	14.3	11.0	Morris, 1988
Girls	16.9	40.9	Morris, 1988

7. INTIMATE PARTNER VIOLENCE

Studies of the frequency of last year intimate partner violence in the Caribbean are summarized in Table 8.6.

Table 8.6. Rates of intimate partner violence in the Caribbean

	COUNTRY	IPV	%	REFERENCE
1	Jamaica	Last year	20	Mc Farlane, 1999
2	Dominican Rep.	Last year	22	Kishor & Johnson, 2004
3	Haiti	Last year	29	Kishor & Johnson, 2004
4	St Croix	Last 2 years	28	Stockman et al., 2014
5	US Virgin Ils	Last 2 years	38	Stockman et al., 2014
6	USA: Black	Last year	11	Cazenaze & Strauss, 1990
7	USA: Black	Last year	23	Caetano et al. 2000

	COUNTRY	IPV	%	REFERENCE
8	USA: White	Last year	3	Cazenaze & Strauss, 1990
9	USA: White	Last year	12	Caetano et al. 2000

Row 1 shows that 20 percent of women aged 15-49 in Jamaica reported having experienced forced sex. Rows 2 and 3 give data for any kind of intimate partner violence. Rows 4 and 5 give data for any kind of intimate partner violence in the last last two years for Blacks in St Croix and the US Virgin Islands. Rows 6 and 7 give data for any kind of intimate partner violence for Blacks in the United States, showing in row 5 about the same rate as in the Caribbean. Rows 8 and 9 give data for any kind of intimate partner violence for Whites in the United States, showing lower rates than for Blacks

8. SEXUALLY TRANSMITTED DISEASES

Rates of sexually transmitted diseases in Jamaica are given in Table 8.7. Row 1 gives results in a 1997 survey of 9,111 of 15-24 year olds in Jamaica showing that 8.5 percent of females and 8.7 percent of males reported an STD in the last year. Row 2 gives results for a sample of women attending a family planning clinic showing that 26 percent had an STD. Rows 3 and 4 give results for a national sample of 958 aged 15-49 years showing that 29 percent of men and 9 percent of women reported a history of STD. Rows 5 and 6 give results in a 1997 survey of 9,111 of 15-24 year olds in Jamaica, showing that 8.5 percent of females and 8.7 percent of males reported an STD in the last year.

Table 8.7   Rates of sexually transmitted diseases in Jamaica (percentages)

	STD	PERCENT	REFERENCE
1	Current	26	Behets et al, 1998
2	Current	19	Dowe et al., 1998
3	Ever: males	29	Jamaica Ministry Health, 2000
4	Ever: females	9	Jamaica Ministry Health, 2000
5	Last year: males	8.7	Norman, 2001
6	Last year: females	8.5	Norman, 2001

9. WORK MOTIVATION AND COMMITMENT

It was noted in Chapter 2 that in the United States Blacks have weaker work motivation and commitment than Whites,and in Chapter 5 that Blacks have weak work motivation and commitment in Sub-Saharan Africa. The weak work motivation and commitment of Blacks was also found in the Caribbean after slavery was abolished in the British colonies in 1833. The British plantation owners found that many of the former African slaves were unwilling to work for wages as day laborers and were unreliable employees. Thomas Atwood, the chief judge in Dominica and later in the Bahamas, described the problem: "It is difficult to make them work: it is sometimes absolutely necessary to have recourse to measures that appear cruel, in order to oblige them to labour" (Fryer, 1984, p. 164). To solve the problem of securing a reliable supply of labor, the plantation owners brought in a number of Indians from the Indian subcontinent and also some Portuguese and Chinese, who were found to be more reliable laborers than the Blacks:

> In Jamaica, Grenada, Guyana, and Trinidad, many ex-slaves rapidly quit the plantations to establish villages where they could live and cultivate subsistence crops without interference from white management and their allies. To replace this lost labor, the planters in these countries used government revenues to import Chinese, Portuguese, and Indians under onerous indentures (Smith, 1984, p. 138).

10. CONCLUSIONS

Blacks in the Caribbean display high rates of psychopathic personality expressed in high rates of crime, sexual precocity, sexually transmitted diseases, intimate partner violence and recklessness in the non-use of contraception, and low rates of moral values, the ability to delay gratification, stable monogamous unions, and work motivation and commitment.

CHAPTER 9

AUSTRALIA

The Aborigines are the indigenous people of Australia. It has been estimated that before the Europeans arrived they numbered around 300,000. Their numbers were considerably reduced following the colonization of Australia by Europeans, partly as a result of diseases contracted from Europeans from which they lacked immunities, and partly as a result of Europeans killing them. In the second half of the twentieth century, the numbers of Aborigines in the censuses of 1961, 1971, and 1981 were recorded as approximately 106,000, 139,000, and 171,000. In 2001 the number had increased to 458,500 and in 2011 it had increased further to 669,736. This rapid increase in numbers has been a result of high birth rates with a TFR (Total fertility Rate) of 2.74, compared with 1.88 for all Australian women, and a reduction of infant and child mortality. The Australian Aborigines have a low average IQ calculated at 62 on the basis of seventeen studies given in Lynn (2016).

1. CONDUCT DISORDER

A high level of psychopathic personality in the Australian Aborigines is expressed among children and young adolescents in their high rates of conduct disorder and of expulsion from schools for psychopathic behavior. In New South Wales in the 1990s, Aborigines comprised three percent of the school population but 12 percent of those suspended and expelled (Gray, Hunter & Schwab, 2000).

2. CRIME

The high prevalence of psychopathic personality disorder in the Australian Aborigines is also expressed in their high rates of crime. Studies documenting the high crime rates of the Australian Aborigines expressed as odds ratios in relation to 1.0 for Europeans are given in Table 9.1. These show incarceration rates of Australian Aborigines for selected years between 1986 and 2011 at between 9.7 and 26.0 times higher than those of Europeans.

Table 9.1. Crime (imprisonment) rates of Australian Aborigines and Europeans (odds ratios)

	YEAR	ABORIGINES	EUROPEANS	REFERENCE
1	1986	9.7	1.0	Cove, 1992
2	1992	26.0	1.0	Broadhurst, 1997
3	1999	14.0	1.0	Australian crime, 2000
4	2004	16.0	1.0	Australian Bureau Statistics, 2005
5	2011	14.6	1.0	Ting, 2011
6	2013	15.0	1.0	Lee, 2013

Row 1 gives data for 1986 showing that Aborigines were over-represented in prison in all the eight Australian states by factors (odds ratios) ranging from 3.0 in the North Territory to 12.5 in South Australia, and by 9.7 for Australia as a whole. Row 2 gives data for 1992 compiled by Broadhurst (1997) reporting that the rate of imprisonment for adults was 26 times higher for Aborigines than for Europeans. He discusses the reasons why Aborigines have so much higher rates of imprisonment. He dismisses the idea that there could be any genetic reasons for this because "the hereditary thesis with its origins in phrenology is now fully discredited" (p. 413). He discusses the possibility that there could be racial bias in police arrests or court convictions, but finds that this is not the case because the police do not arrest Aborigines proportionately more than their crime rates warrant, and because Aborigines report much higher rates of assault by other Aborigines, as compared with Europeans. He concludes that "the key general cause of the disproportionate criminalization of Aborigines is universally perceived to be socioeconomic deprivation and consequential exclusion," and that "the underlying issues of unemployment, poverty, ill-health, dispossession, and disenfranchisement are the causes of the over-involvement of Aborigines in prison" and these are themselves "the product of indirect discrimination." (pp. 453-4) Rows 3, 4, 5 and 6 show that the incarceration rates of Aborigines were consistently 14 to 16 times higher than those of Europeans from 1999 to 2013.

3. SEXUAL PRECOCITY

Greater sexual precocity among Australian Aborigines than among Europeans is shown by the fertility rate of 78 per 1,000 for teenagers aged 15-19 in 2011, compared with 16 per 1,000 of Europeans (Australian Bureau of Statistics, 2011). A study of sexual experience of 1,139 Anglo-Australian and Chinese-Australian 17-20 year old college students is summarized in Table 9.2. For both men and women, significantly higher percentages of Anglo-Australian than of Chinese-Australian students had experience of sexual intercourse.

Table 9.2. College students' experience of sexual intercourse (percentages)

	ANGLO AUSTRALIAN	CHINESE AUSTRALIAN	REFERENCE
Men	67	40	Rosenthal et al., 1990
Women	68	22	Rosenthal et al., 1990

RACE DIFFERENCES IN PSYCHOPATHIC PERSONALITY

4. DRUG AND SUBSTANCE ABUSE

The high prevalence of psychopathic personality in the Aborigines is expressed in their high rates of drug and substance abuse. Results of a number of studies reporting these high rates in Australian Aborigines compared with Europeans are summarized in Table 9.3. Row 1 shows 36 percent of 9-14-year-olds in an Aboriginal community were petrol sniffers. Rows 2 and 3 show much higher rates of hospital admissions for Aborigines for alcoholism and liver cirrhosis in 1977. Rows 4 and 5 show much higher rates of alcohol abuse defined as heavy drinkers in a report suggesting that the "reasons are related back to the sense of powerlessness, low status, and lack of privilege in being a minority."(Callan, 1986, p. 45). Rows 6 and 7 confirm the much higher rates of alcohol abuse by Aboriginal men and women, defined as drinking nine or more standard drinks per drinking session. The high rate of alcohol abuse of Aboriginal women results in a higher rate of fetal alcohol syndrome of 2.76 percent in Aboriginal babies, compared with 0.02 percent in Whites reported by the Human Rights and Equal Opportunity Commission (2005). Rows 8 through 11 show much higher rates of cigarette use by Aboriginals. Rows 12 through 14 show much higher rates of cannabis use by Aboriginal teenagers, and rows 15 and 16 show much higher rates of cannabis use by Aboriginal adults.

Table 9.3. Drug abuse rates of Australian Aborigines and Europeans (percentages)

	DRUG ABUSE	ABORIGINES	EUROPEANS	REFERENCE
1	Petrol sniffing	36	-	Eastwell, 1979
2	Alcoholism	1.58	0.17	Hunt, 1981
3	Liver cirrhosis	0.12	0.04	Hunt, 1981
4	Alcohol: males	30	5	Callan, 1986
5	Alcohol: females	3	1	Callan, 1986

	DRUG ABUSE	ABORIGINES	EUROPEANS	REFERENCE
6	Alcohol: males	53	4	Hunter et al., 1992
7	Alcohol: females	19	0.5	Hunter et al., 1992
8	Cigarettes: males	71	39	Hogg, 1995
9	Cigarettes: females	76	42	Hogg, 1995
10	Cigarettes: males	50	28	Perkins et al., 1994
11	Cigarettes: females	49	20	Perkins et al., 1994
12	Cannabis	29.5	11.9	Gray et al., 1997
13	Cannabis	24	10	Forero et al., 1999
14	Cannabis	11.9	9.0	Zubrick et al., 2005
15	Cannabis: males	66.2	4.9	India et al. 2012
16	Cannabis: females	30.5	2.2	India et al. 2012

In addition to these studies, Perkins, Sanson-Fisher, and Blunden (1994) found that Aborigines were significantly more likely to have used marijuana, heroin, cocaine and petrol sniffing, and Pink & Allbon (2008) have reported that Aborigines were more likely to have used illicit drugs in the past 12 months.

5. INTIMATE PARTNER VIOLENCE

Intimate partner violence between husbands and wives is another characteristic of psychopathic personality for which there is a high prevalence among the Australian Aborigines. Studies summarizing the high rate of violence by Aboriginal men on their female partners are shown in Table 9.4. Row 1 shows that Aboriginal women were ten times more likely to be killed by their men partners than were European women. Row 2 shows the results of a 1994 study in Western Australia that found that Aboriginal women were 45 times more likely to have experienced violence from their husbands than had European women. Row 3 shows over the period 2006-7 Aboriginal and Torres Strait Islander people were 34 times more likely to be hospitalized as a result of domestic violence compared to non-indigenous people, and reports that the true rate of violence in many communities is likely to be higher than that reported (Steering Committee for the Review of Government Service Provision, 2009).

Table 9.4. Intimate partner violence of Australian Aborigines and Europeans (odds ratios)

	IPV	ABORIGINES	EUROPEANS	REFERENCE
1	Homicide	10	1	Cumberworth, 1997
2	Violence	45	1	Donnan, 2001
3	Violence	34	1	Steering ctte, 2009

6. CHILD ABUSE AND NEGLECT

Aboriginal and Torres Strait Islander children have been over-represented in child protection and out-of-home care services compared to other Australian children since the first data were collected in 1990 (Australian Institute of Health and Welfare, 2011). In 2010-2011, 3.46 percent of Aboriginal and Torres Strait Islander children had child protection records of harm or risk of harm from abuse or neglect, compared with 0.45 percent of non-Aboriginal children, indicating that Aboriginal and Torres Strait Islander children were 7.5 times more likely than non-Indigenous children to have experienced harm or risk of harm. The maltreatment most frequently experienced by Aboriginal children was child neglect defined as the failure of the parents to provide for a child's basic needs, including the provision of adequate food, shelter, clothing, supervision, hygiene, or medical attention. Aboriginal and Torres Strait Islander children were also at greater risk than non-Aboriginal children of being sexually abused {sexual abuse}(Steering Committee for the Review of Government Service Provision, 2007).

7. SEXUALLY TRANSMITTED DISEASES

Rates of the sexually transmitted diseases are considerably higher for Aborigines than for Europeans. The rates for gonorrhoea, syphilis and chlamydia for 2009-2011 given by the Australian Bureau of Statistics (2011) are shown in Table 9.5. Rates of HIV are less than 1.0 per 10,000 in both Aborigines and Europeans.

Table 9.5. Sexually transmitted disease rates per 10,000 for Australian Aborigines and Europeans

STD	ABORIGINES	EUROPEANS
Gonorrhoea	109	1.7
Syphilis	2.7	0.47
Chlamydia	164	29.6

8. PROBLEM GAMBLING

Table 9.6 summarizes studies of the greater prevalence of problem gambling in Australian Aborigines than in Europeans. Row 1 gives the percentage of 2.1 of problem gamblers in European Australians in a study carried out in the late 1990s. Row 2 gives a similar percentage (2.5) of problem gamblers in European Australians in the Northern territories in a study carried out in the early 2000s and a rate about three times greater (7.9) in Aborigines. Rows 3 and 4 give two more recent studies reporting a high prevalence of problem gambling in Australian Aborigines.

Table 9.6. Problem gambling in Australian Aborigines and Europeans (percentages)

	ABORIGINES	EUROPEANS	REFERENCE
1	-	2.1	Productivity Commission. 1999.
2	7.9	2.5	Young et al., 2007
3	24.0	-	Stevens & Young, 2009a.
4	13.5	-	Stevens & Young, 2009b

9. WORK MOTIVATION AND COMMITMENT

The low level of work motivation and commitment of the Aborigines living in settlements has been described by the German sociologist Hans Schneider:

> Almost all the inhabitants are unemployed and fully dependent on social security. Any motivation to work has been destroyed by their weekly security cheques. They sit around in a state of boredom and hopelesssness. Faulty machinery is simply left where it breaks down and transistor radios are thrown away when the batteries are flat. Under the supervision of whites they are able to establish a plantation or cattle station and will work there, but as soon as this supervision and instruction is withdrawn the project collapses (Schneider, 1992, p.11).

10. SELF-ESTEEM

A further expression of the high prevalence of psychopathic personality disorder in the Aborigines is their high self-esteem. A study of 195 secondary schools students by Purdie and McCrindle (2002) found that Australian Aborigines had higher self-esteem and a stronger self-concept than a comparison group of 162 Europeans. The Aborigines had a higher rate of endorsement of questions like "I am happy with the sort of person I am [self-acceptance]," "I like the work we do at school [satisfaction with school]," "I get good marks in most of my work at school [academic achievement]," "I have many friends [peer acceptance]," and "I will be successful in what I do when I leave school [career confidence]." These are remarkable results considering the reality of the low levels of achievement of the Aborigines in schools and employment. Similar results however have been found for African-Americans in the United States given in Chapter 2.

11. CONCLUSIONS

In all of the expressions of psychopathic personality (conduct disorders, crime, sexual precocity, child abuse and neglect, drug abuse, intimate partner violence, sexually transmitted diseases, problem gambling and self-esteem) Australian Aborigines have a substantially higher prevalence than Europeans. The Chinese in Australia have lower psychopathic personality than Europeans expressed in their lower prevalence of sexual precocity consistent with other studies showing low psychopathic personality in Northeast Asians reviewed in Chapter 6.

CHAPTER 10

NEW ZEALAND

The Maori are the indigenous peoples of New Zealand, which they settled about 800 AD from Polynesia. Europeans colonized New Zealand in the nineteenth century, and a number of Chinese entered the country in the second half of the twentieth century. In 2001 the Maori were 14 percent of the population, Europeans were 74 percent, and Chinese 6 percent. The Maori were described by an American psychology professor, David Ausubel (1961, pp. 65–73), who made a study of them in the late 1950s and wrote

> Maori parents tend to adopt a passive, uninterested, and *laissez-faire* attitude towards their children's vocational careers. They are more willing to let an adolescent son drift...; they live in an atmosphere of wretched housing and sanitary conditions, uncontrolled drinking, improvident spending, and gross neglect of children...adolescents and adults alike tend to become demoralized, apathetic, and unwilling to take even the simplest steps to improve their lot.

This description suggests they have a high prevalence of psychopathic personality.

1. CONDUCT DISORDERS

Studies summarizing the prevalence of conduct disorders in Asians, Maori, and Europeans are given in Table 10.1. Row 1 gives much

189

higher rates of psychopathic personality in Maori than in Europeans among young adults born in 1977 and followed up in the Christchurch birth cohort study.

Table 10.1 Race differences in conduct disorders (percentages)

	CONDUCT DISORDERS	ASIANS	MAORI	WHITES	REFERENCE
1	Psychopathy	-	19.3	6.4	Marie et al., 2014
2	Conduct disorders	-	11.3	4.0	New Zealand Dept Corrections, 2007
3	Truancy	2.0	5.5	2.1	New Zealand Dept Corrections, 2007
4	School stand-downs	7	58	21	New Zealand Dept Corrections, 2007
5	School suspensions	3	17	4	New Zealand Dept Corrections, 2007
6	Delinquency	-	52	25	Lim et al., 2012

Row 2 gives rates of conduct disorders for 16-18 year olds for 2005, and shows these almost three times greater in Maori than in Whites. Row 3 gives rates of truancy from school as percentages in a sample week in 2004, and shows these approximately fifty percent higher in Maori than in Whites. Row 4 gives rates per 1,000 of school "stand-downs", the New Zealand term for temporary exclusion from school, for 2005, and shows these approximately eight times greater in Maori than in Whites. Row 5 gives rates in 2005 per 1,000 of school suspensions which are longer and can be permanent, and shows these approximately four times greater in Maori than in Whites. Asians have lower rates than Whites for truancy, of school stand-downs and school suspensions. Row 6 gives percentages classified as delinquent in matched samples of youth that sexually offend, and shows that these are approximately twice as great in Maori as in Whites.

2. CRIME

The Maori have much higher rates of crime than Europeans and Asians (Fifield & Donnell, 1980; Lovell & Norris, 1990; Newbold, 2000; Spier, 2001). This is shown in Table 10.2 for criminal convictions for men aged 15 years and older per 1,000 for four categories of crime for the period 1951-1966 (theft includes burglary and fraud, and drunkenness includes vagrancy).

Table 10.2 Race differences in crime (per 1,000)

	CRIME	ASIANS	MAORI	WHITES	REFERENCE
1	Assault	0.42	4.79	0.61	New Zealand Statistics, 1970
2	Theft	0.89	27.57	5.71	New Zealand Statistics, 1970
3	Drunkenness	0.86	15.46	5.17	New Zealand Statistics, 1970
4	Sex crimes	0.13	1.49	0.40	New Zealand Statistics, 1970
5	All crimes	2.30	49.31	11.89	New Zealand Statistics, 1970
6	Violence	-	4.56	2.39	Marie et al., 2014
7	Property	-	3.42	2.44	Marie et al., 2014

The total crime rates for Maori are more than four times higher than those for Whites, while the Asians (ethnic Chinese) had much lower total crime rates about one fifth those of Whites. Rows 6 and 7 give much higher rates of violent and property crime in Maori than in Europeans among young adults born in 1977 and followed up in the Christchurch birth cohort study.

Further evidence for a high Maori crime rate in 1999 was published by Rich (2000) who reported that Maoris were approximately 14 percent of the population but 38 percent of prison admissions. The high Maori crime rate was confirmed in 2007 when it was reported that Maori were 12.5 percent of the population and 50 percent of those in prison (New Zealand Dept of Corrections, 2007) giving Maori a 400 percent over-representation among those committing sufficiently serious crimes to receive a custodial sentence.

A number of social scientists in New Zealand have argued that the higher rate of criminal convictions of the Maori does not reflect a higher rate of offending but is attributable to discrimination, prejudice, and racism of the police and the criminal justice system. This racism and discrimination includes bias in police arrest practices and cultural biases in the justice system that place Maoris at greater risk of being convicted when they appear before the Court (Lovell & Norris, 1990). This explanation has been tested by Fergusson and Horwood and their colleagues in the Christchurch Health and Development Study (CHDS). This is a longitudinal study of a birth cohort of over 1,000 young people that has been studied from birth to age 21. As part of this study, data on both officially recorded convictions and self-reported crime have been gathered at regular intervals. In the first of these studies they examined rates of police contact among young Maoris up to the age of 14. They found that Maoris had rates of police contact that were 2.9 times higher than rates for Europeans, and that the higher rate of crime among the Maori was also present in self-reported offending.

In a second study, Fergusson, Horwood and Swain-Campbell (2003) examined rates of conviction and of self reported serious crimes (assault and burglary) committed between the ages of 18 and 21 by Maori and Europeans in the CHDS cohort. The results are shown in Table 10.3. Row 1 shows that Maori had 5.9 times the rate of conviction for serious crimes as Europeans, somewhat higher than the rate of a little over four times higher rate for all ages shown in Table 10.2. Row 2 gives the Maori self reported crime rate, and shows that this was 3.2 times the rate of Europeans.

Table 10.3. Crime rates per annum of 18-21 year olds in New Zealand (percentages)

	CRIME	MAORI	EUROPEANS	ODDS RATIO
1	Convictions	30.3	5.2	5.9
2	Self-reported	452	136	3.2

The authors of these studies conclude that the results suggest the presence of some biases in the criminal justice system in so far as the conviction rates of Maoris are greater than their self reported crime rates. However, this inference assumes that Maori and European youth are telling the truth when asked about what crimes they have committed. Research in the United States by Huizinga and Elliott (1984) and Hindelang, Hirschi and Weis (1981) has shown that there are large discrepancies between criminal convictions and self-reported crimes, and that Blacks under-report their crimes more than Whites. The most significant results of the Fergusson and Horwood studies are that they show that self-reported crimes by Maori are considerably greater than those of Europeans.

3. RECKLESSNESS IN SEXUAL BEHAVIOR AND SEXUAL PRECOCITY

Recklessness and sexual precocity are two features of psychopathic personality that are captured by rates of teenage pregnancy. Virtually no teenagers want to become pregnant and have babies (Kalmuss,

1992), so teenage pregnancies and births can be regarded as resulting from recklessness in having unprotected sex and sexual precocity. The birth rates of teenage Maori and European women per 1,000 aged 15-17 years are shown for selected years from 1996 to 2003 are given by the New Zealand Department of Corrections (2007), and are shown in Table 10.4. It will be seen that in all three years the rate for Maori teenagers was approximately five times greater than the rate for Europeans.

Table 10.4. Births to teenage women aged 15-17 per 1,000

YEAR	MAORI	EUROPEANS	ODDS RATIO
1996	48.3	9.9	4.8
2000	40.3	7.9	5.1
2003	39.4	7.9	5.0

4. CHILD ABUSE AND NEGLECT

Child abuse and neglect are two further features of psychopathic personality that are expressions of the "inability of function as a responsible parent" in the list of psychopathic characteristics described by the American Psychiatric Association. There are higher rates of these in the Maori. This is shown for children aged 0-16 years for selected years from 1998 to 2003 in Table 10.5 given by the New Zealand Dept of Corrections (2007). It will be seen that in the three years the rate of child abuse and neglect for Maori was between two and two and a half times greater than the rate for Europeans. In another study it was reported that Maori children younger than 5 years were twice as likely to be hospitalized for intentional injury than non-Maori children (Child Youth and Family, 2006).

Table 10.5. Child abuse and neglect per 1,000

YEAR	MAORI	EUROPEANS	ODDS RATIO
1998	12.6	5.0	2.5
2000	12.2	5.2	2.3
2003	12.0	6.0	2.0

5. DRUG USE

In the nineteenth century, the European colonists brought alcohol and introduced the Maori to it. Many of them developed alcohol abuse, and this problem was sufficiently serious in the 1850s that the British colonial authorities prohibited the sale of alcohol to Maori. In the twentieth century, Maori men were 2.7 times more likely to die from excessive alcohol consumption than Europeans, while Maori women were 1.6 times more likely to die from excessive alcohol consumption. The Maori had higher rates of admission to hospitals for liver cirrhosis and diseases of the pancreas resulting from excessive alcohol consumption, and had 1.6 times the rate of drunk-driving accidents than Europeans (Mancall, Robertson & Huriwai, 2000). In a further study, alcohol-related deaths of Maori males were 2.2 times the rate of European males of the same age (Pomare, Keefe-Ormsby, Ormsby & Pearce, 1995). Further studies of the high prevalence of the Maori in substance abuse disorders are given in Table 10.6. Row 1 and show that by the age of 21 Maori had 6.0 times the rate of conviction for marijuana (cannabis) use than Europeans and 5.2 times the rate of self-reported marijuana use. Rows 2 through 4 show that Maori had about 50 percent higher self-reported marijuana use than Europeans. Row 5 shows a higher rate of self-reported cigarette use by Maori than by Europeans. It has also been reported that the Maori are disproportionately over-represented in substance abuse disorders with a 25 percent life time experience (Baxter, 2008).

Table 10.6. Maori and European rates of drug use (percentages)

	DRUG USE	MAORI	EUROPEANS	REFERENCE
1	Marijuana: Convictions	13.1	2.2	Fergusson et al., 2003
2	Marijuana: Self-reported	17.1	3.3	Fergusson et al., 2003
3	Marijuana: Self-reported	20.8	14.0	New Zealand, 2007
4	Marijuana: Self-reported	44.7	32.3	Marie et al., 2014
5	Cigarettes: Self-reported	44.7	32.3	Marie et al., 2014

6. INTIMATE PARTNER VIOLENCE

Psychopathic personality is expressed in violence and the inability to preserve long term loving relationships with spouses and partners. These two characteristics are captured in "intimate partner violence" consisting of assaults by spouses and partners on each other. The lifetime prevalence rates of these for Asians, Europeans and Maori were collected in a national survey of a sample of 4,559 by Morris and Reilly (2003) and the results are given in Table 10.7. It will be seen that the prevalence of violence between intimate partners was nearly twice as frequent among Maori than among Whites, and the use of a weapon was almost three times as frequent. The rates for Asians were the lowest for both measures of intimate partner violence.

Table 10.7. Intimate partner violence (percentages)

	BEHAVIOR	ASIANS	MAORI	WHITES
1	Violence	13.2	31.7	17.3
2	Use weapon	3.4	12.2	4.1

In a later study it was reported that Maori women were significantly more likely to be victims of repeated domestic violence than women from other ethnic groups (Child Youth and Family, 2006).

7. WORK MOTIVATION AND COMMITMENT

The low level of work motivation and commitment of the Maori is indexed by their high rate of welfare dependency reported as 48.3 percent among young adults born in 1977, and followed up in the Christchurch birth cohort study, compared with 24.9 percent in Europeans (Marie et al., 2014).

8. CONCLUSIONS

In all of the expressions of psychopathic personality (conduct disorders, crime, recklessness, sexual precocity, child abuse and neglect, drug use, intimate partner violence and welfare dependency) Maori have a substantially higher prevalence than Europeans. For the three measures for which there are data (conduct disorders, crime and intimate partner violence) Asians, consisting largely of Chinese, have a substantially lower prevalence than Europeans confirming the results showing low psychopathic personality in Northeast Asians reviewed in Chapter 6.

CHAPTER 11

PACIFIC ISLANDERS

The Pacific Islanders are the indigenous peoples of the numerous Pacific islands, the principal of which are Micronesia, Melanesia, Polynesia and Hawaii. These islands were uninhabited by humans until about 6,000–1,000 BC, when they began to be settled by Southeast Asian peoples. It was not until about AD 650 that all the major islands of Polynesia were settled. In classical anthropology the Pacific Islanders were recognized as one of the seven major races by Coon, Garn, and Birdsell (1950). This was confirmed by Cavalli-Sforza, Menozzi, and Piazza (1994) in their genetic classifications, in which Micronesians, Melanesians, and Polynesians appear as a "cluster."

1. CONDUCT DISORDERS

Table 11.1 gives differences in conduct disorders and related conditions of Asians, Hawaiians, and Whites from a study of children aged 6-18 years. Row 1 gives the percentages with externalizing problems (another term for conduct disorders). Rows 2, 3 and 4 give scores for delinquency, aggression, and ADHD. The general pattern of the results is that Asians were significantly lower than Native Hawaiians and Whites on all measures, while Native Hawaiians were slightly but not significantly higher than Whites.

Table 11.1. Race differences in conduct disorders in Hawaii (percentages)

VARIABLE	ASIAN	HAWAIIAN	WHITE	REFERENCE
Externalizing problems	8.96	13.96	13.26	Loo & Rapport, 1989
Delinquency	1.95	3.05	2.57	Loo & Rapport, 1989
Aggression	7.08	10.66	10.54	Loo & Rapport, 1989
ADHD	8.72	11.82	11.19	Loo & Rapport, 1989

2. CRIME

Crime rates in Hawaii for native Hawaiians, Europeans, Northeast Asians (mainly ethnic Japanese), Filipinos, and Puerto Ricans were reported for 1924 and 1930 and are shown in Table 11.2.

Table 11.2. Crime rates in Hawaii per 1,000 population

YEAR	EUROPEANS	FILIPINOS	NATIVE HAWAIIANS	NORTHEAST ASIANS	PUERTO RICANS	REFERENCE
1924	1.5	7.1	3.6	0.8	9.3	Porteus & Babcock, 1926
1930	12.5	16.6	17.0	2.6	28.1	Vernon, 1982

Row 1 gives jail inmates per 1,000 population in 1924 and shows the highest crime rate among Puerto Ricans (largely Blacks) followed by Filipinos and native Hawaiians, with lower crime rates for Europeans,and the lowest rate for Northeast Asians. Row 2 shows the same race differences for the juvenile convictions per 1,000 for the year 1930. The juvenile convictions rates are higher because crime is greater among adolescents, and because the rates for adults are for serious crimes meriting a jail sentence.

Further more recent data for race differences in crime rates in Hawaii are given for 1986 by the Office of the Attorney General (1987), and are shown in Table 11.3. The data are presented for the percentages of the racial groups in the population and their percentages of total arrests (excluding traffic violations), and arrests for murder and drug offenses. The race differences are similar to those found in the first half of the twentieth century shown in Table 11.2. The Northeast Asians have the lowest crime rate. They were 32.5 percent of the population, but only 9 percent of total arrests, 8.1 percent of arrests for murder, and 10.5 percent arrests for drug offenses. Europeans and Filipinos come next for total arrests, which were slightly greater than their percentage in the population. Hawaiians, Puerto Ricans (largely Blacks), and Samoans, are all considerably over-represented among those arrested, in relation to their percentage in the population.

Table 11.3 Crime rates in Hawaii (percentages)

CRIME	EUROPEANS	FILIPINOS	NATIVE HAWAIIANS	NORTHEAST ASIANS	PUERTO RICANS	SAMOANS
Population %	33.9	13.9	12.0	32.5	1.8	1.5
Total arrests	34.5	12.3	23.4	9.0	4.0	4.7
Murder arrests	13.5	54.0	13.5	8.1	-	16.2
Drug arrests	45.5	8.6	16.9	10.5	6.2	2.2

3. ILLEGITIMACY

Race differences in illegitimacy as an index of recklessness in Hawaii for 1988 have been published by the Hawaiian Department of Health. They are expressed as rates per 1,000 live births and are shown in Table 11.4. It will be seen that the Chinese, Japanese, and Koreans had the lowest illegitimacy rates followed by the Europeans. The rates of the Blacks, Filipinos, Portuguese and Samoan are intermediate, while the rates of the Puerto Ricans (largely Blacks) and the native Hawaiians are the highest.

Table 11.4. Illegitimacy rates in Hawaii

BLACK	CHINESE	EUROPEAN	FILIPINO	HAWAIIAN	JAPANESE	KOREAN	PORTUGUESE	PUERTO RICAN	SAMOAN
158	65	137	221	436	112	95	299	434	302

4. SEXUALLY TRANSMITTED DISEASES

Differences in the prevalence of HIV infection in Native Hawaiians and Pacific Islanders and American Europeans are given in Table 11.5, and show the prevalence of infection over twice as great among Native Hawaiians and Pacific Islanders as among Europeans.

Table 11.5 Race differences in the prevalence of HIV per 1,000

HIV	HAWAIIAN-PACIFIC IS	EUROPEANS	REFERENCE
2008	0.21	0.08	Center Disease Control, 2012
2010	0.19	0.09	Center Disease Control, 2012

5. SUBSTANCE ABUSE

The most complete data for alcohol, tobacco, and other drug use among Pacific Islanders in Hawaii have been given by Wong et al.(2004) for a study of approximately 24,000 school students aged 12-18 years carried out in the late 1990s. The results are given in the first six rows of Table 11.6 and show the highest rates among Hawaiians, followed by Whites, and the lowest rates among Asians.

Table 11.6. Race differences in drug and substance abuse (percentages)

DRUG	CHINESE	FILIPINO	HAWAIIAN	JAPANESE	WHITE	REFERENCE
Alcohol abuse	20.9	35.4	51.5	29.6	51.4	Wong et al., 2004
Cigarettes	33.9	64.8	64.2	44.7	58.9	"
Marijuana	15.4	35.4	51.6	27.6	45.8	"
Inhalents	5.3	11.2	11.5	6.1	12.4	"
Cocaine	3.1	4.2	7.2	2.7	6.7	"
Heroine	0.9	1.9	2.7	1.6	2.6	"
Illicit drugs	3.3	8.7	10.6	3.3	8.3	Substance Abuse Admin, 2012

The last row gives elicit substance abuse rates for 2011 for a sample 67,5000 aged 12 plus years obtained from interviews showing similar differences.

6. CHILD ABUSE AND NEGLECT

Dubanoski and Snyder (1980) report that child abuse and neglect are more prevalent in Hawaii among Pacific Islanders (Samoans) who in 1976 were 0.8 percent of the population and 6.6 percent of cases of child abuse and 2.6 percent of cases of neglect. Ethnic Japanese were 27 percent of the population but only 3.5 percent of cases of child abuse and 4.6 percent of cases of neglect.

7. CONCLUSIONS

In all the expressions of psychopathic personality (conduct disorders, crime, illegitimacy, HIV infection, drug use and child abuse and neglect) Pacific Islanders have a higher prevalence than Europeans and ethnic Chinese and Japanese have lower prevalence than Europeans, confirming the low prevalence of psychopathic personality in Northeast Asians documented in Chapter 6.

CHAPTER 12

INUIT

A study of psychopathic personality among the Inuit (formally known as Eskimos) in Alaska was carried out by Murphy (1976). He reported that the Inuit have the word *kunlangeta* for the psychopath, who is described as,

> a man who, for example, repeatedly lies and cheats and steals things and does not go hunting and, when other men are out of the village, takes sexual advantage of many women – someone who does not pay attention to reprimands and who is always being brought to the elders for punishment. When asked what would have happened traditionally to such a person, an Eskimo said that probably 'somebody would have pushed him off the ice when nobody else as looking' (p.1,026).

It might be expected that these informal executions would have reduced the genes for psychopathic personality in the population but the evidence indicates a high level of psychopathic personality.

1. CRIME

Canadian studies and statistics normally give data for Aboriginals who include Native Americans as well as Inuit, and rarely give data disaggregated for the two groups. In the mid-1990s 3 percent of the adult population of Canada were Aboriginal and of these 17 percent of men and 26 percent of women were in prison (Foran, 1995;

Roberts & Melchers, 2003) showing substantially over-representation in prison and higher rates of crime of Native Americans and Inuit combined.

Crime rates in Inuit Nunangat and in the rest of Canada averaged for 2006 to 2008 are given in Statistics Canada (2009), and are shown in Table 12.1 and show considerably higher rates in Inuit apart from robbery (the community of Rigolet, Newfoundland and Labrador, and all communities in Nunavik, Quebec, are excluded from Inuit Nunangat because of police-reported data limitations).

Table 12.1. Crime rates in Inuit Nunangat and in the rest of Canada, per 1,000 population

CRIME	INUIT	REST OF CANADA
Total - all violations	380	60
Total violent incidents	90	10
Homicide	0.3	0
Sexual assault	8	1
Major assault	12	1
Common assault	57	5
Robbery	0.5	0.9
Uttering threats	16	2
Breaking and entering	25	7
Theft	25	22
Mischief	136	11
Disturbing peace	79	3

2. SEXUAL PRECOCITY

Data for 2009 suggesting sexual precocity among Inuit females are given by Statistics Canada (2012a), and are given in Table 12.2. These show the percentages of teenage pregnancy higher in Inuit than in Canada as a whole. These figures may also reflect greater recklessness among the Inuit, because few teenagers become pregnant intentionally.

Table 12.2 Teenage pregnancy (percentages)

POPULATION	UNDER 15 YEARS	15 TO 19 YEARS
Canada	0	4.1
Inuit	0.2	20.0

3. MONOGAMY AND MARRIAGE

There are low rates of stable monogamous unions among Inuit. In 2009, 76.7 percent of Inuit women who gave birth were unmarried, compared with 27.2 percent for Canada as a whole (Statistics Canada, 2012b).

4. DRUG AND SUBSTANCE USE

Muckle et al. (2011, p.1081) have written that, "Nowadays, alcohol, smoking and drug use are major public health and social concerns in Canadian aboriginal groups." Studies confirming this are summarized in Table 12.3. Row 1 gives results for the use of marijuana (cannabis), an illegal drug in Canada, for women over 15 years of age reporting having used the drug in the 12 preceding months showing a much greater prevalence among the Inuit. Rows 2 through 4 give results for ever having used marijuana, solvents, and cocaine for 248 Inuit women in Arctic Quebec. Row 5 gives results for binge drinking in women defined as consuming five or more drinks when they drank for a sample of Inuit in Arctic Quebec showing this was five to six times more prevalent than among women from the general Canadian population. Row 6 gives results for cigarettes smoking for women showing this was more than twice as high among Inuit as among Whites. Rows 7 through 9 show high percentages of a sample of 215 Inuit women smoking cigarettes, taking marijuana, and binge drinking during pregnancy. Row 10 shows a much lower percentage of marijuana use in a sample of 15 year olds European Canadians.

Table 12.3 Drug and substance use (percentages)

	DRUG	INUIT	WHITE	REFERENCE
1	Cannabis	47	5	Muckle et al., 2011
2	Marijuana	81	-	Muckle et al., 2011
3	Solvents	62	-	Muckle et al., 2011
4	Cocaine	23	-	Muckle et al., 2011
5	Binge drinking	62	11	Muckle et al., 2011
6	Cigarettes	73	33	Muckle et al., 2011
7	Cigarettes	90	-	Fraser et al, 2012
8	Binge drinking	38	-	Fraser et al, 2012
9	Marijuana	36	-	Fraser et al, 2012
10	Marijuana	-	14	Adlaf et al., 2004

5. IRRESPONSIBLE PARENTING

Irresponsible parenting measured by child maltreatment, abuse, and neglect is more prevalent in Aboriginal children, who include Native Americans as well as Inuit, than among Europeans. Thus, 5 percent of Canadians aged 15 years and younger were classified as Aboriginal in the 1996 Canadian Census, yet "Aboriginal children, meaning children of Inuit, Métis, or First Nations ancestry, represent up to 40 percent of the 76,000 children and youth placed in out-of-home care in Canada" because of parental neglect and maltreatment (Blackstock et al., 2004, p.901). In a later review of the literature, Fluke et al. (2013, p. 47) have written:

The chronic over-representation of Aboriginals in Canadian child welfare care has been well documented. Analysis based on national census data reported that while 5% of children in Canada were Aboriginal in 1998, Aboriginal children made up 17% of children reported to the child welfare authorities, 22% of substantiated reports of child maltreatment, and 25% of children placed in care in Canada

6. CONCLUSIONS

In all of the expressions of psychopathic personality (crime, sexual precocity, teenage pregnancy, illegitimacy, drug use and child maltreatment, abuse and neglect) Inuit in Canada have a substantially higher prevalence than Europeans.

CHAPTER 13

LATIN AMERICA

The racial composition of the countries of Latin America varies considerably from 85 percent European in Argentina to 1 percent European in Honduras. The non-European populations are largely Native American Indian and Mestizo, except in Brazil where approximately 53 percent are European, 15 percent are Native American Indian and Mestizo, while 6 percent are Black and 22 percent are Mulatto.

1. CRIME

Crime rates for homicide, rape and assault per 100,000 population for ten Latin American countries for the years 1993-1996 have been given by Rushton and Templer (2009). Their results are given in Table 13.1, together with the medians of the rates averaged for 36 European countries. It will be seen that for both crimes the rates in the Latin American countries are generally higher than in Europe. The rates for homicide are the most accurate and are higher in all the Latin American countries than in Europe and in most cases much higher. The medians for the Latin American countries are given in the last row and show that the median for homicide is more than three times higher than in Europe, the median for rape one and a half times higher, and the median for assault almost twice as high. These results suggest that psychopathic personality is higher in Latin American than in European populations.

Table 13.1. Crime rates per 100,000 population for 10 Latin American countries and Europe

	WHITE %	HOMICIDE	RAPE	ASSAULT
Europe	100	5	6	43
Argentina	85	16	9	22
Belize	4	24	15	288
Chile	52	6	10	106
Colombia	20	66	4	80
Ecuador	5	23	9	39
Guyana	2	18	18	1111
Honduras	1	60	1	54
Panama	10	17	4	20
Paraguay	7	15	4	79
Venezuela	42	23	17	148
Median	-	17	9	80

Further data confirming these figures for the rates for homicide were given as 21 per 100,000 population for the 26 Latin American countries for the years 2003-2008 compared with 5 per 100,000 population in 42 European countries by Harrendorf, Heiskanen, and Malby (2010).

Race differences in crime in Brazil are summarized in Table 13.2. Row 1 gives the percentages of Asians, Blacks, Mulattoes, and Whites in the population in 2000. Row 2 gives rates of conviction for homicide for 2003 showing that Asians have the lowest rate at 0.4 percent drawn from 1 percent of the population; Whites also have a relatively low homicide rate at 39.7 percent drawn from 53 percent of the population. Mulattoes have a higher homicide rate at 49.9 percent for 40 percent of the population, while Blacks have the highest homicide rate at 9.8 percent drawn from 6 percent of the population. Row 3 gives crime rates for 2008 using the criterion for the classification by race of IBGE (Brazil Goverment Institute for Statistics), in which Blacks and Mulattoes are aggregated into the one category shown in the table as Mulattoes. The results show that Mulattoes and Blacks were 46 percent of the Brazil population and 55.2 percent of prison population, while Whites were 53 percent of the population but only 37.5 of the prison population. Row 4 shows that in 2013 the Brazile Ministry of Justice estimated that Black (with Mulattoes and Africans combined) accounted for some 75 percent of the prison population.

Table 13.2. Race differences in crime in Brazil (percentages)

	ASIAN	BLACK	MULATTO	WHITE	REFERENCE
Population percent	1	6	40	53	Lopes, 2006
Homicide percent	0.4	9.8	49.9	39.7	Lopes, 2006
Crime percent	-	-	55.2	37.5	IBGE, 2008
Prisoners: 2010	-	-	75	25	Rede Justiça Criminal (2016)

2. MORAL VALUES

The Defining Issues Test was developed by Rest (1979, 1986) and consists of a number of stories in which the leading actor is confronted with a moral dilemma. The problem is to discern the moral principle involved and choose the course of action consistent with it. This test has been given to high school students in several countries with different racial populations and the results for Belize expressed as d in relation to American scores of zero are shown in Table 13.3. The population of Belize is approximately 44 percent Mestizo, 30 percent Mulatto and 4 percent White (Phillips, 1996). The high school students in Belize obtained lower scores than Americans indicating a poorer understanding of moral values in this predominantly Mestizo and Mulatto population.

Table 13.3. Moral values assessed by the Defining Issues Test (ds)

LOCATION	MESTIZO/MULATTO	WHITE	REFERENCE
Belize	-0.48	0.00	Rest, 1986

3. SEXUAL PRECOCITY

Studies providing data on age of first sexual intercourse of boys in Mexico city and Guatemala city, and for Black and White boys in the United States in the 1980s, are summarized in Table 13.4. The population of Mexico is approximately 30 percent Native American Indian, 60 percent Mestizo and 9 percent White, and the population of Guatemala is approximately 55 percent Native American Indian, 42 percent Mestizo, and 3 percent White (Philips, 1996). Rows 1 and 2 give age of first sexual intercourse of boys in Mexico city and Guatemala city, and Rows 3 and 4 give age of first sexual intercourse of Black and White boys in the United States. The results suggest greater sexual precocity in the predominantly Native American Indian and Mestizo populations, and in American Blacks, than in American Whites.

Table 13.4. Age of first sexual intercourse of boys

	COUNTRY	AGE	REFERENCE
1	Mexico City	15.7	Morris, 1988
2	Guatemala City	14.8	Morris, 1988
3	USA: Blacks	14.4	Zelnik & Shah, 1983
4	USA: Whites	15.9	Zelnik & Shah, 1983

4. RECKLESSNESS: NON-USE OF CONTRACEPTION

Studies providing data for percentages that used contraception during first sexual intercourse in a number of Latin American countries in the 1980s have been given by Morris (1988), and are summarized in Table 13.5, all showing the low percentages that used contraception.

Table 13.5. Use of contraception (percentages)

COUNTRY	SEX	AGE	% USE CONTRACEPTION
Mexico City	F	17.0	22.3
Mexico City	M	15.7	30.7
Guatemala City	F	16.7	10.4
Guatemala City	M	14.8	14.9
Panama	F	16.7	11.3
Brazil	F	16.6	14.9
Paraguay	F	16.9	12.2

A study in Brazil gives data for the use and non-use of contraception on the occasion of first sexual intercourse of women age 15-19 in the northeast where the population is largely Mulatto and Black, and in the southeast where the population is largely White (Gupta, 2000). The results are shown in Table 13.6. It will be seen that in 1986 and again in 1996, greater percentages in the northeast did not use contraception on the occasion of first sexual intercourse.

Table 13.6. Race differences in the non-use of contraception in first sex in Brazil (percentages)

	YEAR	NORTHEAST	SOUTHEAST
1	1986	94	76
2	1996	79	52

5. INTIMATE PARTNER VIOLENCE

Studies of the frequency of last year intimate partner violence in Latin American countries are summarized in Table 13.7. These percentages are much higher than those of around 3 to 12 percent for White women in the United States.

Table 13.7. Intimate partner violence in Latin America

	COUNTRY	IPV	%	REFERENCE
1	Colombia	Ever	44	Kishor & Johnson, 2004
2	Colombia	Ever	40	Friedemann-Sanchez, 2012
3	Colombia	Last 12 months	22	Friedemann-Sanchez, 2012
4	Nicaragua	Ever	52	Ellsberg et al., 1999

	COUNTRY	IPV	%	REFERENCE
5	Nicaragua	Ever	30	Kishor & Johnson, 2004
6	Peru	Ever	42	Kishor & Johnson, 2004
7	Peru	Ever	61	WHO, 2002

6. DRUG ABUSE

Race differences in a 2000 survey of 9,633 of postpartum women in Rio de Janeiro reporting whose who had smoked cigarettes while pregnant are given by Leal (2006) and are shown in Table 13.8. It will be seen that Blacks had the highest percentage of smokers, followed by mulattoes, and Whites the lowest.

Table 13.8. Race differences in smoking in Brazil

BLACK	MULATTO	WHITE
18.5	14.9	10.3

7. SEXUALLY TRANSMITTED DISEASES

Rates of the sexually transmitted diseases are given in Table 13.9 for the results in a 2000 survey of 9,633 of postpartum women in Rio de Janeiro who had babies that were syphilitic, showing that Blacks had the highest percentage of syphilitic babies, followed by mulattoes, and Whites the lowest.

Table 13.9. Race differences in sexually transmitted diseases (percentages)

STD	BLACK	MULATTO	WHITE	REFERENCE
Baby syphilitic	3.0	1.9	0.8	Leal, 2006

8. WORK MOTIVATION AND COMMITMENT

It was noted in Chapter 2 that in the United States Blacks have weaker work motivation and commitment than do Whites, and in Chapters 5 and 8 that in Africa and the Caribbean that Blacks have weaker work motivation and commitment than do Indians. The weak work motivation and commitment of Blacks was also found in Brazil. Large numbers of Blacks were imported into Brazil from the sixteenth to the nineteenth centuries as slaves to work in sugar and tobacco plantations. After slavery was abolished in Brazil in 1888 it was found that they were too unreliable and unsuitable for employment for wages. The European plantation owners solved this problem by bringing in Japanese as indentured workers (Halpern, 2004).

9. NATIVE AMERICAN INDIANS

We saw in Chapter 2 that Native American Indians in the United States have a high level of psychopathic personality. It has been proposed by Raine (2013, pp.20-21) that this is true of the Mundurucu and the Yanomamö, tribes of Native American Indians in the Amazon basin. He describes the Mundurucu as "fiercely aggressive head hunters" with "features of psychopathic behaviour", and he describes the Yanomanö as "fearless and highly aggressive" and possessing "precisely the features of Western psychopaths."

10. CONCLUSIONS

The studies reviewed in this chapter show a high level of psychopathic personality in the Mestizo and Mulatto populations of Latin America than of Europeans expressed in high rates of crime, weak moral values, sexual precocity, intimate partner violence, drug abuse, recklessness in the non-use of contraception, sexually transmitted diseases and poor work motivation and commitment.

CHAPTER 14

DIFFERENCES IN PSYCHOPATHIC PERSONALITY ACROSS NATIONS

Hitherto we have examined race differences in psychopathic personality within nations and geographical regions. The general pattern of the racial and ethnic differences in the measures and expressions of psychopathic personality shows a considerable degree of consistency. Sub-Saharan Africans and Native Americans almost invariably show high levels of psychopathic personality. Hispanics typically appear intermediate between high scoring Sub-Saharan Africans and Native Americans and lower scoring Europeans. South Asians and North Africans from the Indian sub-continent and in Europe, Canada, and South Africa typically show slightly higher psychopathic personality than Europeans. Northeast Asians, principally Chinese and Japanese, almost invariably show the lowest level of psychopathic personality compared with all other peoples. Australian Aborigines show a very high levels of psychopathic personality. New Zealand Maori and other Pacific Islanders show higher levels of psychopathic personality than Europeans. We consider now how far these differences within counties and geographical regions are consistent with differences in psychopathic personality across nations.

1. EPIDEMIOLOGICAL STUDIES

We examine first epidemiological studies of the prevalence of psychopathic personality in a number of countries. One of the most satisfactory of these is the international personality disorder study. This study used the same diagnostic interview procedure to assess the prevalence of psychiatric patients with definite and probable anti-social personality disorder in twelve cities in a "standardized and reliable way" (Lorager, Janca & Satorius, 1997). The results for London and New York are discarded because they are multi-racial cities in which only approximately half the population is indigenous British in London and American in New York. The results are given in Table 14.1.

Table 14.1. Prevalence of definite and probable anti-social personality disorder (percentages)

COUNTRY	DEFINITE	PROBABLE
Austria (Vienna)	4.0	10.0
Germany (Munich)	5.3	8.0
India (Bangalore)	6.4	6.4
Japan (Tokyo)	0.0	1.8
Kenya (Nairobi)	16.0	22.0
Luxembourg	1.9	1.9
Netherlands (Leiden)	3.1	4.6
Norway (Oslo)	10.4	18.8
Switzerland (Geneva)	0.0	3.1
UK (Nottingham)	6.0	18.0
Europe (6 cities)	4.5	9.1

The data show that the one Northeast Asian country (Japan) has the lowest prevalence of both definite and probable anti-social personality disorder. The one Sub-Saharan African country (Kenya) has the highest prevalence of definite and probable anti-social personality disorder. The prevalence of anti-social personality disorder in European peoples is best represented as the average of the seven cities (Vienna, Munich, Luxembourg, Leiden, Oslo, Geneva and Nottingham) to give the results shown in the bottom row. The results for the one South Asian country (India) show higher definite and lower probable anti-social personality disorder than in the average of the European counties. Taken as a whole, the results show that the prevalence of anti-social personality disorder is lowest in the Northeast Asians represented by Japan, intermediate in the Europeans (seven cities) and South Asians represented by India, and highest in sub-Saharan Africans represented by Kenya.

These results are confirmed by six major community studies of the prevalence of anti-social personality in the general population that are summarized in Table 14.2, showing sample sizes and lower rates in the three Northeast Asian countries than in the three European countries.

Table 14.2. Prevalence of anti-social personality (percentages)

COUNTRY	N	PREVALENCE	REFERENCE
Hong Kong	7,229	1.65	Chen et al., 1993
South Korea	5,100	1.7	Lee et al.,1990
Taiwan	11,004	0.08	Hwu et al., 1989
Canada	3,258	3.7	Bland et al., 1988
New Zealand	1,498	3.1	Wells et al.,1989
USA	18,571	2.6	Robins et al., 1991

2. HOMICIDE

We examine next the prevalence of homicide as the most reliable measure of crime rates and as an expression of psychopathic personality across countries. Most homicides are recorded and consequently provide a reasonably accurate measure of their prevalence. National data for rates of homicide per 100,000 population in the early twenty-first century have been reported by Harrendorf, Heiskanen and Malby (2010) and are given in Appendix 1. These have been classified by race, and are shown in Table 14.3.

Table 14.3. Race differences in rates of homicide per 100,000 population

RACE	COUNTRIES	HOMICIDE
Sub-Saharan Africa	50	27
Latin America	26	21
Caribbean	24	16
Southeast Asia	11	11
Central Asia	5	6
South Asia	9	5
Europe	42	5
North Africa	7	4
Northeast Asia	5	2
Pacific Islands	25	2

These figures largely confirm the race differences within countries. The highest rates of homicide are present in the 50 Sub-Saharan African countries (27 per 100,000 population). The next highest (21 per 100,000 population) are in the 26 Latin American countries largely attributable to their large Native American Indian populations and, in Brazil, sub-Saharan Africans. These followed by the 24 Caribbean countries (16 per 100,000 population) attributable to their large sub-Saharan African populations. After these come substantially lower rates in 11 Southeast Asian countries (11 per 100,000 population). These are followed by a considerable further drop to Central Asia, (6 per 100,000 population), South Asia (5 per 100,000 population), Europe (5 per 100,000 population) and North Africa (4 per 100,000 population). The consistency of the figures for these four populations is attributable to these being all Caucasoid peoples. The two lowest rates are for the five Northeast Asian countries (2 per 100,000 population) and 25 Pacific Islands countries (2 per 100,000 population). The low rates for the Northeast Asian countries are consistent with their low rates of psychopathic personality within countries and with the results shown in Tables 14.1 and 14.2. The low rates for the Pacific Island countries are inconsistent with their rates of psychopathic personality within countries; this is possibly attributable to the under-recording of homicide in a number of these countries.

3. CRIME

Here we adopt crime rates as a measure of race differences in psychopathic personality following Cooke, Mitchie, Hart and Clark (2005): "psychopathy is recognized as an important predictor of criminal behaviour." We compare crime rates in Europeans and in non-Europeans in the countries in which two or more races are present. These within-country differences are calculated as odds ratios with Europeans set at 1.0 and are given in Table 14.4. For example, the Aborigines compared with Europeans in Australia are over-represented in crime rates by a factor of 16.1 and Maori in New Zealand are over-represented by a factor of 5.9. Where a race is present in more than one country, as is the case with sub-Saharan Africans, all the comparisons are given and are averaged. These results show that the Australian Aborigines have the highest level of psychopathic personality (16.1) followed by Sub-Saharan Africans (6.9), New Zealand Maori (5.9) and Native Americans (2.2). South Asians have the same rates as Europeans (1.0) and Northeast Asians (0.3) have a lower rates than Europeans.

Table 14.4. Race differences in rates of crime within countries
(Odds ratios, Europeans set at 1.0)

RACE	COUNTRY	ODDS RATIOS
Aborigines	Australia	16.0
Sub-Saharan Africans	Multiple	6.9
Sub-Saharan Africans	South Africa	6.0
Sub-Saharan Africans	United States	7.5
Sub-Saharan Africans	Britain	6.2
Sub-Saharan Africans	France	8.1
Maori	New Zealand	5.9
Native Americans	United States	2.2
South Asians	Britain	1.0
Northeast Asians	Multiple	0.5
Northeast Asians	United States	0.3
Northeast Asians	Britain	0.7

4. CORRUPTION

Here we examine corruption as another measure of racial differences in psychopathic personality across countries. The prevalence of corruption in 174 countries for 2012 has been measured by the Corruption Perceptions Index published by Transparency International (2013). This defines corruption as the abuse of public office for private gain and measures the degree of corruption perceived to exist in public officials and politicians. The Corruption Perceptions Index (CPI) gives a countries' scores between 1 and 100 with a high score indicating a low level of perceived corruption. The scores of the countries are given in Appendix 2. These have been categorized by the racial composition of the populations for those countries in which one race comprises the majority of the population. Countries in which no one race comprises the majority of the population are excluded from this analysis, as with homicide rates given in section 2.

The results are given in Table 14.5. We see that the greatest corruption is present in the seven central Asian countries (24.7) followed by the 42 sub-Saharan African countries (32.4).

Table 14.5. Race differences in rates of corruption (CPI); high scores = low corruption)

RACE	N COUNTRIES	PI
Central Asian	7	4.7
Sub-Saharan African	42	2.4
Pacific Islander	4	3.5
Native American	12	3.7
South Asian	28	8.2
Southeast Asian	6	9.1
Caribbean	7	2.0
Northeast Asian	7	7.4
European	50	3.6

High levels of corruption are also present in the four Pacific Islander countries (33.5) and in the 12 Native American countries (33.7). The least corruption is present in the 50 European countries (63.6) followed by the seven Northeast Asian countries (57.4). The score for the Northeast Asian countries is reduced by the very low score of 8 for North Korea. When this is excluded, the score for the other six is 65.6, making this group the least corrupt. The high score for the seven Caribbean countries seems anomalous in comparison with the low score for the Sub-Saharan African countries. The likely explanation for this is that Europeans and mixed race mulattoes are the principal holders of public office in Caribbean countries.

5. LIFE HISTORY

Rushton (2000) advanced the theory that there are race differences in r-K life history. His r-K life history theory was drawn from biology, where species are categorized on a continuum running from r strategists to K strategists; r strategists have large numbers of offspring and invest relatively little in them, while K strategists have fewer offspring and invest heavily in them by feeding and protecting them during infancy and until they are old enough to look after themselves. Fish, amphibians and reptiles are r- strategists (having large numbers of offspring and minimum investment) while mammals are K strategists (having fewer offspring and greater investment). The K strategy is particularly highly evolved in monkeys, apes and humans. Species that are K strategists have a syndrome of characteristics of which some of the most important are larger brain size, higher intelligence, longer gestation, and a slower rate of maturation in infancy and childhood.

It is proposed here that some of Rushton's r-K-life history differences are attributable to differences in psychopathic personality. Among these are that r-strategists display early sexuality and teenage childbearing, sexual promiscuity, impulsivity, high aggressiveness, high self-esteem, high rates of crime, high rates of sexually transmitted diseases and a preference for instant gratification. These are all features of psychopathic personality and justify the identification of psychopathic personality with these features of r- life history.

Rushton proposed that there are racial differences r-K life history. He applied this theory to the three major races of *Homo sapiens*: Mongoloids, Caucasoids (Europeans, South Asians and North

Africans), and Negroids (Sub-Saharan Africans) and proposed that Mongoloids are the most *K* evolved and Negroids the least *K* evolved, while Caucasoids fall intermediate between the two although closer to Mongoloids. Rushton supported his theory by documenting that the three races differ in a large number of r-*K* life history characteristics. Though a number of exceptions have been highlighted (see Dutton, 2018) in most instances these three racial groups different in the predicted direction of Blacks being fasted, Mongoloids being slowest, and Caucasians being intermediate.

Rushton's theory has been extended by Meisenberg and Woodley (2013) who have constructed a measure of the r-K personality dimension for societies. The variables they use for r-selection are (1) teenage pregnancy rates (proportion of children born to women aged 19 and below) as a measure of early sexuality; (2) STDs (sexually transmitted diseases including syphilis, gonorrhea and clamydia, but excluding HIV/AIDS because of its recent African origin) as a measure of strong and promiscuous sexuality together with a disregard for the future consequences of heedless behaviour; (3) homicide rates; and (4) crime rates (measured as crime victimization obtained by the Gallup World Poll) as measures of impulsive and uncontrolled aggression. The measures they use for K-selection are (1) percentage use of contraception among married couples as a measure of low fertility and (2) savings rates (gross domestic savings rate average 1975-2005) as a measure of a strong capacity for the postponement of instant gratification. They combine these six variables to construct a measure of Rushton's r-K personality dimension for each of 161 countries and present the scores for the major races with the mean for Europeans set at 100.0 (Sd = 15). Their results are given in Table 14.6. These results confirm Rushton's theory that Northeast Asians are the most K evolved followed by Europeans, and sub-Saharan Africans the least K evolved, and that Europeans are closer to Northeast Asians than to Sub-Saharan Africans. They also extend Rushton's theory by the addition of Middle Easterners and Southeast Asians who fall between Europeans and sub-Saharan Africans, and Native Americans who appear as the least K evolved.

Table 14.6. Race differences in K personality (Meisenberg &
Woodley, 2013)

RACE	K PERSONALITY
Northeast Asian	104.7
European	100.0
Middle Eastern	89.4
Southeast Asian	86.3
Sub-Saharan African	74.2
Native American	70.7

Rushton's r-K theory of race differences has been extended further by Minkov (2014) who has measured race differences in K from the World Values Survey. He has found a "national K factor" in these data measured principally by the importance parents attach to their children possessing the negative psychopathic personality characters of thrift, obedience and responsibility. His results giving scores classified by race are shown in Table 14.7. His scores for 71 nations are given in Appendix 3. It will be seen that these results provide further confirmation of Rushton's theory that Northeast Asians are the most K evolved followed by Southeast Asians, Europeans and South Asians, while Sub-Saharan Africans are the least K evolved. Caribbeans and Latin Americans (excluding Argentinians and Uruguayans) scored the second least K evolved, as would be expected of populations with large numbers of native American, Mestizzo and Negroid populations.

Table 14.7. Race differences in Life History Strategy

RACE	N COUNTRIES	LH
Northeast Asian	5	93
Southeast Asian	5	79
European	30	18
South Asian	12	-20
Caribbean & Latin American	9	-41
Sub-Saharan African	10	117

The conclusions reached by Meisenberg and Woodley shown in Table 14.6 and by Minkov (2014) shown in Table 14.7 are largely consistent showing Northeast Asians the most K evolved, Europeans high K evolved and sub-Saharan Africans low K evolved. There is a minor inconsistency in that Meisenberg and Woodley find Europeans higher K than Southeast Asians while Minkov finds Southeast Asians higher K than Europeans. Evidently, then, Rushton's Differential K is a fruitful theory and recent attempts have been made to develop it in order to make sense of the anomalies already noted (see Figueredo et al., 2017 or Dutton, 2018).

6. CONCLUSIONS

The data on race differences in the epidemiology of anti-social personality disorder and in rates of homicide, crime, corruption and in r-K life history summarized in this chapter are generally consistent with those of the within-country studies given in previous chapters. The results confirm that the Northeast Asians have the lowest psychopathic personality indexed by low rates of anti-social personality, homicide, crime and corruption and low K life history.

The next lowest rates of psychopathic personality are in the Europeans. The reverse pattern of indices of high psychopathic personality is typically present in sub-Saharan Africans, Native Americans, Latin Americans with their large Native American and sub-Saharan African (in Brazil) populations, and by Caribbeans with their largely Sub-Saharan African populations. Intermediate values for these indicators of psychopathic personality are typically present in Pacific Islanders, Southeast Asians, South Asians and North Africans.

CHAPTER 15

GENETICS OF RACE DIFFERENCES IN PSYCHOPATHIC PERSONALITY

In this penultimate chapter we consider how far these race differences in psychopathic personality have a genetic basis.

1. HERITABILITY OF PSYCHOPATHIC PERSONALITY

Numerous studies have shown that crime and psychopathic personality have a significant heritability. The first of these was carried out in Denmark by Johannes Lange (1931) on the criminality of identical and fraternal twins. He found 30 pairs of same sex twins in whom at least one had a criminal record. There were 13 identical twins, among whom both twins had criminal records in 10 pairs. There were 17 fraternal twins, among whom both twins had criminal records in two pairs. There was therefore a much greater similarity between the identical twins indicating a significant genetic influence on criminality. Subsequent reviews of the heritability of crime by Eysenck and Gudjonsson (1989), Raine (1993), Lykken (1995),

237

and Viding, Fontaine, and Larson (2013) have all estimated the heritability of crime at about 0.5.

A large number of twin studies have confirmed that psychopathic personality has a moderately high heritability and is therefore significantly genetically determined. Mason & Frick (1994) summarize eight studies of the similarity for psychopathic personality of identical and same sex fraternal twins, in all of which identical twins showed greater similarity than fraternals, and which taken together show a heritability of 0.41. Three further twin studies of psychopathic personality summarized by Nigg & Goldsmith (1994) produced a heritability of 0.56. A more recent study reports a correlation on the Psychopathic Personality Inventory of 0.46 for identical twins and a zero correlation for fraternal twins indicating a 0.46 heritability (Blonigen, Carlson, Krueger & Patrick, 2003). An analysis of identical twins reared apart gives an estimate of the heritability of psychopathic personality of 0.41 for children and 0.28 for adults (Segal, 2012, p.140). The significant heritability of psychopathic personality has been confirmed by adoption studies {adoption studies}showing that adopted people who are psychopathic have a greater percentage of psychopathic biological relatives than are present in the general population (4 percent compared with 1 percent), indicating genetic transmission of psychopathic personality (Schulsinger, 1972). A more recent twin study estimated the heritability of psychopathic personality at 69 percent while non-shared environmental influences explained 31 percent (Tuvblad et al., 2014).

It has also been found that a number of the expressions of psychopathic personality have a significant genetic determination. The heritability of impulsivity was estimated at 0.40 and risk taking at 0.36 from data on identical twins reared apart analyzed by Hur and Bouchard (1997). Silberg et al. (1996) estimate the broad heritability of hyperactivity-conduct disorder at 0.88. Dunne, Martin, Statham et al. (1997) calculate the heritability of age of first sexual intercourse at 0.72 for men and 0.49 percent for women and Segal (2012) analyzing identical twins reared apart data estimates it at 0.39 for both men and for women. The heritability of drug abuse and dependence is estimated at 0.45 from identical twins reared apart data (Segal, 2012).

In a review of the genetics of psychopathic personality, Siever and Kuluva (2012, p. 71) have concluded that "it is fairly well accepted that multiple genes, in combination with other factors, contribute to the development of personality disorder."

These heritability studies showing that psychopathic personality has a moderately strong genetic basis among individuals make it probable that the race differences in psychopathic personality also have a genetic basis. This case is further strengthened by two considerations. First, a study of Black children adopted by White families found that in late adolescence they were more impulsive, extroverted, aggressive, rebellious, and hedonistic than matched Whites (DeBerry, 1991). Second, there are consistent race differences in psychopathic personality in a variety of geographical locations throughout the world. This consistency is particularly striking in Sub-Saharan Africans who have a high level of psychopathic personality in Africa, the Caribbean, the United States, Britain, Canada, France and the Netherlands.

2.NEURO-ANATOMY AND NEURO-PHYSIOLOGY OF PSYCHOPATHIC PERSONALITY

There is substantial evidence that psychopathic personality has a neurological basis. This lies principally in the frontal lobes which have the function of inhibiting psychopathic personality and behavior. Damage to the frontal lobes typically impairs their inhibiting function and results in loss of control of psychopathic behavior and an increase in psychopathic personality (Perez, 2012). Psychopathic behavior is associated with a low volume of the frontal lobes suggesting that these are not exerting their inhibiting function effectively (Gregory, Ffytch, Simmons et al., 2012; Raine, 2013). Hyperactivity of the amygdala has also been found to be associated with aggressive conduct disorders in adolescents (Passamonti, Fairchild, Goodyer et al., 2010) and increases in the aggressive behavior associated with psychopathic personality (Siever & Kuluva, 2012).

3. PSYCHOPATHIC PERSONALITY AND LOW ANXIETY

Cleckley (1976, p. 257) stated that "psychopaths are very sharply characterized by the lack of anxiety," and numerous studies have confirmed this (Lykken, 1995; Skeem, Poythress, Edens et al., 2003; Corr, 2010). The effect of the low level of anxiety is that the conditioned anxieties of social disapproval and punishment that inhibit psychopathic behavior in normal individuals are not acquired so readily in psychopaths. This has been shown by Gao, Raine, Venables, Dawson & Mednick (2010) in a study finding that poor electrodermal anxiety conditioning assessed in a sample of 1,795 children at age 3 was significantly associated with criminal offending at age 23. The authors conclude that "individuals are hypothesized to learn to avoid psychopathic and criminal acts by successfully associating stimuli that are associated with psychopathic events with later socializing punishments. This association learning is hypothesized to result in an increase in anxiety and anticipatory fear whenever the individual contemplates the commission of an psychopathic act, which in turn motivates the individual to avoid such stimuli and the commission of psychopathic, rule-breaking behavior… poor fear conditioning at age 3 predisposes to crime at age 23. Poor fear conditioning early in life implicates amygdala and ventral prefrontal cortex dysfunction and a lack of fear of socializing punishments in children who grow up to become criminals."

These studies showing that low anxiety is a determinant of psychopathic personality raise the question of whether there may be race differences in anxiety that contribute to the differences in psychopathic personality. There is positive evidence for this summarized in studies from the United States in Table 15.1. Row 1 gives results showing the highest prevalence of anxiety in Asians and lowest in Blacks, with Whites intermediate. Row 2 shows that in a sample of young adults life the prevalence of anxiety disorders in Blacks was approximately half that in Whites, with Hispanics intermediate but closer to Whites.

Table 15.1. Race differences in anxiety disorders (percentages)

CONDITION	ASIAN	BLACK	HISPANIC	WHITE	REFERENCE
1 General anxiety disorder	7.3	4.5	-	6.8	Scott et al., 2002
2 Any anxiety disorders	-	3.2	5.9	6.3	Turner & Lloyd, 2004
3 Any anxiety disorders	-	24.7	-	29.1	Breslau et al., 2005
4 Any anxiety disorders	-	23.8	-	29.4	Breslau et al., 2006
5 Panic attacks	-	3.1	-	4.9	Breslau et al., 2006
6 Social phobia	-	10.8	-	12.6	Breslau et al., 2006
7 General anxiety disorder	-	1.37	-	2.71	Hinle etal., 2009
8 General anxiety disorder	2.6	2.4	6.2	7.1	Woodward et al., 2012

Row 3 gives results of the National Comorbidity Survey (NCS; Kessler et al., 1994), a nationally representative survey of 8,098 15 to 54 years olds, and shows the lifetime prevalence rate of anxiety disorders in Blacks lower than that in Whites. The rates in this study are higher than those given in row 2 because the sample were older and had more time in which to experience an anxiety disorder. Rows 4 through 6 give results of the follow-up to the National Comorbidity Survey, the National Comorbidity Survey–Replication (NCS-R), consisting of a national sample of 9,282 aged 18 years old or older, and show that Blacks have a lower lifetime risk than

Whites for the development of any anxiety disorder, panic attacks, and social anxiety disorder. Row 7 gives further results showing that Blacks have lower prevalence of anxiety disorder than Whites. Row 8 gives the highest prevalence of anxiety disorder for Whites followed by Hispanics and Asians, and the lowest prevalence in Blacks. The Asians include south and south east Asians which may explain their anomalously low prevalence compared with the result in row 1.

In addition to these American studies, low levels on anxiety have also been reported in Africa. A study of college students in South Africa has shown that Blacks have a lower score than Whites on neuroticism, a measure of anxiety with which it is correlated at 0.86 (McCrae, 2002, p.111). Low prevalence rates of 0.3 percent of social anxiety in Nigeria have been reported by Gurege et al. (2006) and of 1.9 percent in South Africa by Williams et al.(2008) compared with 7.5 percent in the United States (Kessler et al.,1994). Higher levels of anxiety in Chinese and Japanese than of Whites have been reported in Hawaii by Austin and Chorpita (2004).

The low levels of anxiety of Blacks are confirmed by their low rates of suicide. Studies showing this have been reviewed by Lester (1998) and are summarized in Table 15.2.

Table 15.2. Racial differences in suicide per 100,000

	COUNTRY	YEAR	BLACK	COLORED	SOUTH ASIAN	AMER-INDIAN	WHITE
1	S. Africa	1984	3.0	4.6	9.9	-	18.4
2	USA	1950	1.0	-	-	-	2.4
3	USA	1980	6.1	-	-	13.3	13.2
4	Zambia	1967-71	6.2	-	-	-	20.8

In all of the four studies the suicide rate for Blacks is substantially lower than that of Whites. In a more recent study in South Africa the suicide rate of Blacks was approximately half that of whites (Burrows, 2007).

Anxiety has a significant heritability calculated at .54 from identical twins reared apart by Segal (2012, p.188). Putting this evidence together suggests that genetically based race differences in anxiety contribute to differences in poor fear conditioning and these contribute to differences in psychopathic personality.

4. GENETICS OF PSYCHOPATHIC PERSONALITY

A number of studies have suggested that a genetic basis of psychopathic personality is a low level of the MAOA gene which produces the enzyme monoamine oxidase A. This was first identified by Bruner, Nelen, Breakfield, Ropers and van Oost (1993) in study of psychopathic men in an extended family the Netherlands, all of whom lacked the MAOA gene. This theory was supported by the discovery by Cases, Seif, Grimsby et al. (1995) that knocking out the MAOA gene in mice makes them highly aggressive. Further support for the theory came from the discovery by Caspi, McClay, Moffit et al (2002) that low levels of MAOA are associated with aggressive behavior in children, and from the discovery by Eisenbrger, Way,Taylor et al (2007) that low levels of MAOA are associated with aggressive behavior in men and women. These results have been further confirmed by Beaver and his colleagues who have shown that psychopathic personality is associated with the 2-repeat allele of the MAOA gene (Beaver, Wright, Boutwell, Barnes, DeLisi and Vaughn, 2013). This allele is present in 5.5 percent of Black males and 0.1 percent of Caucasian males. In Black males those possessing the allele had higher psychopathic scores, a significantly greater history of committing acts of serious violence, and of being imprisoned. In further studies, Beaver, Wright, Boutwell et al. (2013) have shown that possession of the 2-repeat allele of the MAOA gene is associated with arrests, incarceration and lifetime psychopathic behavior, and

Beaver, Barnes and Boutwell (2013) have shown that possession of this allele confers an increased risk for young males for shooting and stabbing people. They reported that those with the 2-repeat allele had a 0.50 probability of having shot or stabbed someone during the last year, while for those without the 2-repeat allele the probability was 0.07. Thus, those with the allele had a 12.9 times greater probability of exhibiting this expression of psychopathic personality. They report that this is allele is carried by 5.2 percent of African-American men, 0.1 percent of Caucasian men and 0.00067 percent of Asian men. Similar race differences have also been found by Reti, Xu, Yanofski, McKibben et al. (2011), who reported 4.7 percent of African-American men and 0.5 percent of Caucasian men carried the allele.

The possible contribution of the MAOA gene to race differences in psychopathic personality is further suggested by the discovery by Lee and Chambers (2007) that the Maori of New Zealand have substantially lower levels of the MAOA gene than Caucasians, consistent with their higher levels of psychopathic personality. In more recent work, the T allele of the MAOA gene has been identified with low activity and is higher among Africans (88 percent), intermediate in Europeans (71 percent), and lowest in East Asians (40 percent) (MAOA, 2014).

CHAPTER 16

THE EVOLUTION OF RACE DIFFERENCES IN PSYCHOPATHIC PERSONALITY

In this final chapter we consider how the race differences in psychopathic personality evolved. In 2002 I proposed that the cold winters and springs of Eurasia exerted selection pressures against psychopathic personality and for an enhancement of pro-social personality in the European and especially the Northeast Asian peoples. This theory has been endorsed and elaborated by Temper (2013) and Nyborg (2013), and is further elaborated here.

1. SELECTION PRESSURES FOR PRO-SOCIAL PERSONALITY

It is proposed that the cold winters and springs of Eurasia exerted four selection pressures for an enhancement of pro-social personality and against psychopathic personality in the European and especially the Northeast Asian peoples. These were for the evolution of stronger male-female pair bonding, for an increased capacity to delay gratification, and for a greater need to maintain harmonious and co-operative social relations. We consider these in turn. First, the weakness of strong male-female pair-bonding based on love is a central component of the psychopathic personality. Male-female pair bonding based on love is not present in the non-human apes (Dunbar, 2010) and evolved in *Homo sapiens*, but we have seen throughout this book it has evolved to different degrees in the races such that is weakly present in sub-Saharan Africans, and increasingly strongly present in South Asians, Europeans and Northeast Asians. To explain this, we follow Quinlan (2008) in believing that stronger male-female bonding based on love evolved as a result of the need for both parents to provide care for their children. This need was relatively weak as a selection pressure for male-female bonding in equatorial sub-Saharan Africa because in that benign climate women could feed their children throughout the year by gathering plant and insect foods with little or no help from men.

Male-female bonding based on love evolved more strongly in the Caucasoids, and even more strongly in the Northeast Asian Mongoloids, because of the need for co-operation between parents for provisioning children to survive during the cold winters of Eurasia. Plant and insect foods were not available for much of the year, especially in the winter and spring. During these seasons, women and children needed men to provide them with meat foods that they obtained through hunting. These men would have been the fathers of their children, and would have been required to provide long-term commitment to their female mates and children, and to engage in more responsible and concerned parenting. They did this because they were strongly pair-bonded with their women partners in relationships based on love.

The second selection pressure exerted by cold winters and springs of Eurasia for an enhancement of pro-social personality would have been an increased capacity to delay gratification by collecting and storing food for future consumption. This point has been made by Ainslie (2013): "If we take the main challenge of climate to be cold, there is a simple hypothesis that makes this challenge a factor in both economic and cultural advance: A climate that is too cold to grow crops for part of the year demands foresight and self-control skills, which then serve as resources for other development." While plant and insect foods were available throughout the year in sub-Saharan Africa and there was no need to store them, many foods in Eurasia were only available at certain times of the year and these had to be stored for the future. Plant foods were generally only available in the summer and fall, and some potential animal foods were only available at particular times. For instance, salmon enter rivers to spawn for a few weeks each year, during which large numbers can easily be killed and stored either by smoking or freezing. Some herd mammals such as reindeer migrate in large herds for a few weeks each year when they can be killed and stored for future consumption. To take advantage of these opportunities required foresight and co-operation between group members.

The third selection pressure of cold winters and springs for an enhancement of pro-social personality would have been that as men became increasingly reliant on group hunting they had to develop a greater capacity for co-operation, the maintenance of harmonious social relations, and stronger control over aggression towards other men. Thus "large game hunting was a crucial resource and harvesting such game depended on co-operation and coordination among hunting parties." (Sterelny, 2013, p.88).

The fourth selection pressure of cold winters and springs for reduction of psychopathic personality would have been that effective co-operative hunting would have required a reduction of promiscuous sexuality, cheating and other forms of psychopathic behavior that disrupted harmonious and co-operative relations within groups of men. As Tomasello et al. (2012) have put it: "Human cognition and sociality thus became ever more collaborative and altruistic as human individuals became ever more interdependent." A variant of this thesis has been advanced by Wilson (2014), who argues in his book *Does Altruism Exist?* that altruism evolved as the result of group selection in populations that depended on co-operative hunting, i.e. the European and especially the Northeast Asian peoples in the northern latitudes.

All of the components of psychopathic personality would have been disadvantageous for survival during the cold winters and springs of Eurasia. The colder the winters and springs, the stronger the selection pressures would have been for an enhancement of pro-social personality and against psychopathic personality. The selection pressure against psychopathic personality would have been weakest on Australian Aborigines and Sub-Saharan Africans, a little stronger on Pacific Islanders, somewhat stronger on South Asians, North Africans and Native American Indians, still stronger on Europeans, and strongest on Northeast Asians who were exposed to successively harsher winters and springs and hence to increases in pro-social personality and reductions in psychopathic personality.

2. REDUCTION IN TESTOSTERONE

The principal neurophysiological adaptation by which a reduction in psychopathic personality evolved most in the Caucasoid and Northeast Asian peoples was the reduction of the male hormone testosterone. Numerous studies reviewed by Siever and Kuluva (2012) have reported that there are high concentrations of testosterone present in those with psychopathic personality and that these are associated with aggressive behavior (Book, Starzyk & Quinsey, 2001; Brooks & Reddon, 1996; Dabbs, 2000), and with crime, deviance, homicide, and attention deficit hyperactivity disorder (Gladue, 1991; Dabbs, 2000; Mazur & Booth, 2014; Sher, 2014, Tremblay et al., 1998), with early sexual intercourse in adolescent males and females (Halpern, Udry & Suchindran, 1997, 1998), with financial risk taking (Sapienza, (Maestripieri & Zingales, 2009), and with pathological gambling (Blanco, Ibanez, Blanco-Jerez & Baca-Garcia, 2001). For instance, a study of 4,462 male veterans found that those with high testosterone were more delinquent in childhood and criminal as adults, had more drug use, and more sexual partners (Dabbs & Morris, 1990). A study of 306 university students found that those with high testosterone were more aggressive (Harris, Rushton, Hampson & Jackson, 1996). A study of Black boys aged 5-11 years with conduct disorders, and of the same age without conduct disorders, showed that those with conduct disorders had higher testosterone levels (Dabbs, 2000). It has also been shown that exposure to fetal testosterone is associated with self-reported measures of offending (Hoskin & Ellis, 2015).

Race differences in testosterone levels showing that these are higher in Blacks than in Whites have been reported in several studies. Ross, Bernstein, Judd, Pike and Henderson (1986) reported that testosterone levels were 19 percent higher in 50 Black male college students than with matched Whites. Ellis and Nyborg (1992) reported that testosterone levels were 3 percent higher in older Blacks than in Whites. It has been shown in a sample without prostate cancer that Blacks in the United States had higher serum testosterone levels than Whites (Blacks: 424T 30ng/di, n= 126; Whites: 380T 19ng/di, n= 126) (Kubricht, Williams, Whatley, Pinckard & Eastham, 1999). A study of race differences in salivary testosterone levels (pmol) has shown that these are highest in Blacks in the Congo (259, n=33) than in Europeans in the United States (259, n=106) (Ellison, Bribiescas, Bentley & Campbell, 2002).

Evidence for race differences in testosterone levels are supported by differences in prostate cancer which is largely caused by high levels of testosterone and has a higher prevalence in Blacks and a lower prevalence in Northeast Asians, than in Whites (Polednak, 1989). The incidence and mortality rates for the major races are given in Table 16.1. Rows 1 and 2 show the incidence and mortality rates per 100,000 in the USA are highest in Blacks followed by Whites and Hispanics, and lowest in Asians. Row 3 gives age-adjusted incidence (new cases) per 100,000 for Koreans and Whites in California 1988-92 showing a much lower rate in Koreans. Row 4 gives age-adjusted incidence for Chinese and Japanese, Blacks and Whites per 100,000 in the USA 1975-96, showing a much lower rate in the Northeast Asians and a higher incidence in Blacks. Rows 5 and 6 show life time risk and death rates in England 2008–2010 are lowest for Asians, intermediate in Whites, and highest in Blacks.

Table 16.1. Race differences in prostate cancer

MEASURE	ASIANS NORTHEAST	ASIANS	BLACKS	HISPANICS	WHITES	REFERENCES
Incidence	-	91.3	220.3	106.7	153.5	Wingo et al.,1998
Mortality	-	11.1	66.0	16.6	24.1	Wingo et al., 1998
Incidence	17.2	-	-	-	101.0	Gomez et al., 2003
Incidence	33.5	-	137.0	-	101.0	Mosli, 2003
Lifetime risk	-	7.9	29.3	-	13.2	Lloyd et al., 2015
Death	-	2.3	8.7	-	4.2	Lloyd et al., 2015

One of the effects of the reduction of testosterone in South Asians, Europeans and Northeast Asians is that this delays their maturation. This has been shown for a number of characteristics. It has been shown that in Britain the age of voice breaking in boys at puberty is earliest in sub-Saharan Africans, later in South Asians, and latest in Europeans (Piffer, 2011). Dutton, Van der Linden and Lynn (2016) have drawn upon a variety of national level markers of testosterone and divided up the nations according to dominant racial group. They have shown that in many cases the differences are in the direction of Blacks having the highest testosterone levels, East Asians the lowest and Whites intermediate.

3. CO-EVOLUTION OF RACE DIFFERENCES IN PSYCHOPATHIC PERSONALITY AND INTELLIGENCE

There is a negative relationship across races in psychopathic personality and the differences in intelligence, such that the races with higher intelligence have lower psychopathic personality. This negative association is shown in Table 16.2 which gives the race differences in psychopathic personality given in Table 14.4 and the differences in intelligence documented in Lynn (2006, 2015). It will be seen that the Australian Aborigines have the highest rate of psychopathic personality and the lowest intelligence, followed by the sub-Saharan Africans with the second highest rate of psychopathic personality and the second lowest intelligence. New Zealand Maori, Native Americans and South Asians come intermediate on both measures, while Europeans and finally Northeast Asians and have the lowest rates of psychopathic personality and the highest intelligence.

Table 16.2. Race differences in IQ and rates of crime (odds ratios, Europeans set at 1.0)

RACE	IQ	P
Australian Aborigines	62	6.0
Sub-Saharan Africans	70	7.5
New Zealand Maori	84	5.9
Native Americans -USA	86	2.2
South Asians – Britain	92	1.0
Europeans	100	1.0
Northeast Asians–Britain	105	0.7

The explanation for this negative relationship between race differences in psychopathic personality and intelligence is that these both evolved as adaptations to the environmental conditions in which the races evolved. As explained above, these consisted of the severity of the winter and spring that required reductions in psychopathic personality and greater intelligence for survival. The Northeast Asians were exposed to the most severe winters and springs and hence evolved the lowest psychopathic personality and the highest intelligence, followed successively by the Europeans, North Africans and South Asians, New Zealand Maori, Sub-Saharan Africans, and Australian Aborigines, resulting in the negative relation between psychopathic personality disorder and intelligence across the major races.

This does not imply that psychopathic personality disorder and intelligence are associated at the level of individuals for which the evidence is inconclusive. The race differences in psychopathic personality and intelligence evolved through different processes. The differences in psychopathic personality evolved through a reduction in testosterone in the Caucasoids and Mongoloids, as noted above, while the differences in intelligence evolved through improvements in neurological processing and increases in brain size documented in Lynn (2006, 2015).

4. CONTRIBUTIONS OF INTELLIGENCE AND PSYCHOPATHIC PERSONALITY TO SOCIAL PATHOLOGIES

We began this study by noting that Herrnstein & Murray (1994) in their book *The Bell Curve* raised the problem that while differences in intelligence can explain some portion of the differences between Blacks, Hispanics and Whites in the United States in a number of important social pathologies including crime, unemployment, poverty, illegitimacy, welfare dependence, low rates of marriage and low birth weight babies, differences in intelligence cannot explain the totality of these differences. They showed that when Blacks, Hispanics, and Whites are matched for intelligence and age, the differences in these social pathologies were reduced but they were not entirely removed.

Herrnstein and Murray's analysis of this problem is summarized in Table 16.3. This sets out for a number of these social pathologies the percentages of Blacks, Hispanics, and Whites showing the characteristic, followed by the percentages of the three groups showing the characteristic when they are matched for the same age of 29 years and the same IQ of 100.

Table 16.3. Racial and ethnic differences before and after matching for age and IQ (percentages)

MEASURE	BLACKS	HISPANICS	WHITES
Crime	13	6	2
Crime: matched for IQ	5	3	1
Marriage	54	76	78
Marriage: matched for IQ	58	75	79
Unemployment	21	14	10
Unemployment: matched for IQ	15	11	11
Illegitimacy	21	14	10
Illegitimacy: matched for IQ	15	11	11
Poverty	26	18	7
Poverty: matched for IQ	9	11	6
Welfare	49	30	13

RACE DIFFERENCES IN PSYCHOPATHIC PERSONALITY

Welfare: matched for IQ	30	15	12
Low birthweight babies	10	5	3
Low birthweight babies: matched for IQ	6	5	3

It will be seen that for all of the social pathologies, matching the racial and ethnic groups for age and IQ reduces the disparities to some degree but in no case is the disparity eliminated. One of the objectives of this monograph has been to examine whether the additional factor that Herrnstein and Murray were looking for to explain the residual disparities consists of the racial and ethnic differences in psychopathic personality. The first three of the pathologies set out in Table 16.3, consisting of rates of crime, marriage and unemployment, have already been considered as expressions of psychopathic personality. We see from the data set out in the table that, when Blacks and Hispanics are matched with Whites for age and IQ, differences in crime rates are considerably reduced but remain substantial, with Black rates still 250 percent greater than White and Hispanic rates 150 percent greater than White. For rates of marriage, matching for age and IQ makes very little difference to the differences between Blacks, Hispanics, and Whites. We can conclude that the differences in marriage rates are almost entirely a function of differences in psychopathic personality. For male unemployment, matching for age and IQ reduces the differences between Blacks and Whites substantially, but Blacks still have unemployment rates 36 percent greater than Whites. With regard to Hispanics, however, matching for age and IQ eliminates the difference from Whites.

The remaining four variables in Table 16.3 can be regarded as secondary effects of racial and ethnic differences in psychopathic personality. For women's illegitimacy rates, matching Blacks to Whites for age and IQ reduces illegitimacy rates only a little, and the illegitimacy rate of Blacks remains approximately five times greater than that of Whites. With regard to Hispanics, matching with Whites produces a greater reduction in illegitimacy rates but they remain substantially greater than those of Whites. These residual racial and ethnic differences in women's illegitimacy rates can be understood as partly determined by differences in psychopathic personality, which contribute to an early onset of sexual activity and the non-use of contraception, which in turn contribute to illegitimacy rates. With regard to women on welfare, we see that matching for age and IQ reduces the Black–White difference by 50 percent and the Hispanic-White difference by 84 percent. The residual differences can be understood as partly determined by differences in psychopathic personality producing higher rates of illegitimate children, and single motherhood among Blacks and Hispanics, which is a major factor responsible for becoming dependent on welfare.

With regard to poverty, matching for age and IQ reduces the Black–White difference by 77 percent, and the Hispanic–White difference by 74 percent. Nevertheless the poverty rate of Blacks remains 50 percent above that of Whites{, and that of Hispanics remains a little greater. Poverty rates are to some degree determined by differences in unemployment, participation in the labor force, and numbers of illegitimate children, which are themselves partially determined by differences in psychopathic personality.

The final variable in Table 16.3 consists of low birth weight babies defined as those weighing less than 5.5 pounds at birth. Matching for age and IQ reduces the Black–White disparity by a little more than half but has no effect on the Hispanic–White disparity. After this matching, Blacks continue to have twice as many low birth weight babies as Whites while Hispanics have two thirds more than Whites. These residual differences are likely to be due, to some degree, to Black and Hispanic women behaving less responsibly during pregnancy because of poor nutrition, alcohol and drug abuse, and can be interpreted as secondary effects of differences in psychopathic personality.

When we look at the general pattern of the data set out in Table 16.3, we note that after matching for age and IQ, Hispanics fall between Whites and Blacks for all the social pathologies except for poverty. This pattern is consistent with the racial and ethnic differences in psychopathic personality for which we have seen Hispanics generally fall between Whites and Blacks. This strengthens the case that the residual group differences in these social pathologies are understandable in terms of differences in psychopathic personality. The exception of the greater rate of poverty for Hispanics than for Blacks is probably largely due to their greater average numbers of children. Our final conclusion is therefore that differences in intelligence and in psychopathic personality make independent contributions of about the same magnitude to the racial and ethnic differences in the social pathologies documented by Herrnstein and Murray.

REFERENCES

Abma, J., Driscoll, A. & Moore, K. (1998). Women's degree of control over first sexual intercourse. *Family Planning Perspectives*, 30, 12-18.

Abrahams, N., Jewkes, R., Hoffman, M. & Laubsher, R. (2004). *Bulletin of the World Health Organization*, 82, 330-337.

Achenbach, T. M. (1992). *Manual for the child behavior checklist*. Burlington, VT: University of Vermont.

Adlaf, E. M., Begin, P. & Sawka, E. (2004). *Canadian Addiction Survey (CAS): a national survey of Canadians' use alcohol and other drugs: prevalence of use and related harms: detailed report*. Ottawa: Canadian Centre on Substance Abuse, 2005.

Aghaei, A. & Golparvar, M. (2014). Prevalence of personality disorders symptoms among male high school students in Isfahan, Iran. *European Journal of Psychology and Educational Studies*, 1, 22-29.

Ainslie, G. (2013). Cold climates demand more intertemporal self-control than warm climates. *Behavioral & Brain Sciences*, 36, 481-482.

Akhtar, R., Ahmetoglu, G. & Chamarro-Premuzic, T. (2013). Greed is good? Assessing the relationship between entrepreneurship and subclinical psychopathy. *Personality and Individual Differences*, 54, 420-425.

Alink, L. R. A., Euser, S., van Ijzendoorn, M. H. & Bakermans-Kranenburg, M. J. (2013). Is elevated risk of child maltreatment in immigrant families associated with socioeconomic status? Evidence from three sources. *International Journal of Psychology*, 48, 117-127.

Alegria, A. A., Petry, N.M., Hasin, D. S., Liu, S. M., Grant, B.F . & Blanco, C. (2009). Disordered gambling among racial and ethnic groups in the US: results from the national epidemiologic survey on alcohol and related conditions. *CNS Spectrums* 14,132–142.

Allen, J. C., Briskman, J., Humayun, S. et al. (2013). Heartless and cunning. Intelligence in adolescents with antisocial behavior and psychopathic traits *Psychiatry Research*, 210, 1147-1154.

Amar, A. R. (1998). Three cheers and two quibbles for Professor Kennedy. *Harvard Law Review*, 111, 1256–1269.

Amaro, H., Jaj, A., Vega, R. R. et al. (2001). Racial and ethnic disparities in the HIV and substance abuse epidemics. *Public Health Reports*, 116, 434-448.

Ambugo, E. A. (2014). Cross-country variation in the sociodemographic factors associated with major depressive episode in Norway, the United Kingdom, Ghana, and Kenya. *Social Science and Medicine,* 113, 154-160.

American Association for Protecting Children (1988). *Highlights of official child neglect and abuse reporting.* Denver, CO: American Humane Association.

American Psychiatric Association (1994, 2000). *Diagnostic and statistical manual of mental disorders.* Washington, D.C.: American Psychiatric Association.

Anderson, E. (1980). Some observations on black youth unemployment. In B. E. Anderson & I. V. Sawhill (Eds) *Youth employment and public policy.* Englewood Cliffs, NJ: Prentice-Hall.

Andretta, J. R., Woodland, M. H., Ramirez, A. M., & Barnes, M. E. (2013). ADHD symptom frequency and ADHD symptom count clustering in African-American adolescents with juvenile court contact. *Journal of Forensic Psychiatry and Psychology,* 24, 570-593.

APA (American Psychological Association) Zero Tolerance Task Force (2008). Are zero tolerance policies effective in the schools? *American Psychologist*, 63, 852-862.

Arbisi, P. A., Ben-Porath, Y. S. & McNulty, J. (2002). A comparison of MMPI-2 validity in African-American and Caucasian psychiatric patients. *Psychological Assessment*, 14, 3-15.

Archer, R. P. (1997). *The MMPI-A: assessing adolescent psychopathology.* Mahwah, NJ: Lawrence Erlbaum.

Ards, S. D., Myers, S. L., Malkis,A., Sugmeg, E. & Zhu, I. (2003).

Racial disproportionality in reported and substantiated child abuse and neglect: an examination of systematic bias. *Children and Youth Services Review*, 25, 375-392.

Argo, T. & Black, D. W. (2004). The characteristics of pathological gambling. In J. E. Grant & M. N. Potenza (Eds) *Understanding and Treating Pathological Gambling*. Washington DC: American Psychiatric Association.

Arnold, L. E., Elliott, M., Sachs, L. et al. (2003). Effects of ethnicity on treatment attendance, stimulant response/dose and 14-month outcome in ADHD. *Journal of Consulting and Clinical Psychology*, 71, 713-727.

Austin, A. A. & Chorpita, B. F. (2004). Temperament, anxiety, and depression: Comparisons across five ethnic groups of children. *Journal of Clinical Child and Adolescent Psychology*, 33, 216-226.

Australian Bureau of Statistics (2011). *Australian Demographic Statistics*. Canberra: Australian Govt.

Australian Crime (2000). *Facts and Figures*. Australian Institute of Criminology.

Australian Institute of Health and Welfare (2011). *Child Protection*. Canberra: Australian Govt

Ausubel, D.P. (1961). *Maori Youth*. New York: Holt, Rinehart & Winston.

Australian Bureau of Statistics (2005). *Population by age and sex, Australian states and territories*. Canberra: Australian Bureau of Statistics Cat. No. 3201.0.

Bachanas, P. J., Morris, M. K., Lewis-Gess, J. K. et al. (2002). Psychological adjustment, substance use, HIV knowledge, and risky sexual behavior in at risk minority females: developmental differences during adolescence. *Journal of Pediatric Psychology*, 27, 373-384.

Bachman, J. G., O'Malley, P. M., Freedman-Doan, P., Trzesniewski, K. H. & Donnellan, M. B. (2010). Adolescent self-esteem: Differences by race/ethnicity, gender, and age. *Self & Identity*, 10, 445-473

Bachman, R. (1992). *Death and violence on the reservation: homicide, family violence and suicide in American Indian populations.* Westport, CT: Auburn House.

Bachrach, C. A., Stolley, K. S. & London, K. A. (1992). Relinquishment of premarital births. *Family Planning Perspectives*, 24, 27–32.

Backman, J. G. (1970). *Youth in transition.* Ann Arbor, MI: Institute for Social Research.

Banfield, E. (1974). *The unheavenly city revisited.* Boston: Little, Brown.

Bankston, C.L. & Zhu, M. (2002). Being well vs doing well: Self-esteem and school performance among immirrant and non-immigrant racial and ethnic groups. *International Migration Review*, 36, 389-415.

Barbé, B. (1951). *Médicine Tropicale*, 11, 33-50.

Barkley, R. A. (1997). Behavioral inhibition, sustained attention, and executive functions: constructing a unifying theory of ADHD. *Psychological Bulletin*, 121, 65–94.

Barry, D. T., Stefanovics, E. A., Desai, R. A., Potenza, M. N. (2011a). Differences in the associations between gambling problem severity and psychiatric disorders among Black and White adults: findings from thenational epidemiological survey on alcohol and related conditions. *American Journal on Addictions*, 20, 69-77.

Barry, D. T., Stefanovics, E. A., Desai, R. A., Potenza, M. N. (2011b). Gambling problem severity and psychiatric disorder among Hispanic and white adults: findings from a nationally representative sample. *Journal of Psychiatric Research*, 45, 404- 414.

Bates, L.A. (2013). Does it matter if teachers and schools match the student? *Social Science Research*, 42, 1180-1190.

Baxter, J. (2008). *Maori mental health needs profile: A review of the evidence.* Palmerston North, New Zealand: Te Rau Matatini.

Beaver, K. M., Barnes, J. C. & Boutwell, B. B. (2013). The 2-repeat allele of the MAOA gene confers an increased risk for shooting and stabbing behaviors. *Psychiatric Quarterly*, 84, DOI 10.1007/s11126-

013-9287-x

Beaver, K. M., Wright, J.P., Boutwell, B. B., Barnes, J. C., DeLisi, M. & Vaughn, M. G. (2013). Exploring the association between the 2-repeat allele of the MAOA gene promoter polymorphism and psychopathic personality traits, arrests, incarceration, and lifetime psychopathic behavior. *Personality and Individual Differences*, 54, 164-168.

Behets, F.M., Ward, E. Fox, L. et al. (1998). Sexually transmitted diseases are common in women attending Jamaican family planning clinics. *Sexually Transmitted Information*, 74 (Suppl), 123-127.

Bell, N. S., Amoroso, P. J., Yore, M. M., Smith, G. S. & Jones, B. H. (2000). Self-reported risk-taking behavior s and hospitalization for motor vehicle injury among active duty personnel. *American Journal of Preventative Medicine*, 18, 85-95.

Berman, M. & Leask, L. (1994). Violent death in Alaska: who is most likely to die? *Alaska Review of Economic and Social Conditions*, 29, 1–12.

Bernstein, D. P., Cohen, P., Velez, C. N., Schwab-Stone, M., Siever, L. J., & Shinsato, L. (1993). Prevalence and stability of the DSM-III—R personality disorders in a community-based survey of adolescents. *American Journal of Psychiatry*, 150, 1237-1243.

Bernstein, D. P., Cohen, P., Skodal, A., Bezirganian, M. D. & Brook, J. S. (1996). Childhood antecedents of adolescent personality disorder. *American Journal of Psychiatry*, 153, 907–913.

Berrington, A. (1996). Marriage patterns and inter-ethnic unions. In D. Coleman & D. Salt (Eds) *Ethnicity in the 1991 Census*. London: HMSO.

Bertrand, J. T., Makani, B., Hassig, S. E. et al. (1991). AIDS-related knowledge, sexual behavior and condo use among men and women in Kinshasha, Zaire. *American Journal of Public Health*, 81, 53-58.

Bickel, F. & Qualls, R. (1980). The impact of school climate on suspension rates in the Jefferson county public schools. *The Urban Review*, 12, 79–86.

Biello, K. B., Ickovics, J., Niccolai, L., Lin, H. & Kershaw, T. (2013). Racial differences in age at first sexual intercourse: residential racial segregation and the Black-White disparity among U.S. adolescents. *Public Health Reports,* 128 (Suppl 1), 23-32.

Binson, D., Dolcini, M. M., Pollack, L. M. & Catania, J. A. (1993). Multiple sexual partners among young adults in high risk cities. *Family Planning Perspectives,* 25, 268-273.

Birthrong, A. & Latzman, R.D. (2014). Aspects of impulsivity are differentially associated with risky sexual behaviors. *Personality and Individual Differences,* 57, 8-13.

Blackburn, R. (1998). Psychopathy and personality disorder: implications for interpersonal theory. In J. D. Cook, A. E. Forth & R. D. Hare (Eds) *Psychopathy: Theory, Research and Implications for Society.* Dortrecht, Netherlands: Kluwer Academic.

Blanc, A. K. & Rutenberg, N. (1993). Coitus and contraception: the utility of data on sexual intercourse for family planning programs. *Studies in Family Planning,* 22, 162-175.

Blackstock, C., Trocmé, N. & Bennett, M. (2004). Child maltreatment investigations among Aboriginal and non-Aboriginal families in Canada. *Violence Against Women,* 10, 901-916.

Blanco, C., Ibanez, A., Blanco-Jerez, C.R. & Baca-Garcia, L. (2001). Plasma testosterone and gambling. *Psychiatric Research,* 105, 117-121.

Blanchard, E. B., Bassett, J. E. & Koshland, E. (1977). Psychopathy and delay of gratification. *Criminal Justice and Behaviour,* 4, 265-271.

Bland, R. C., Orn, H. & Newman, S. C. (1988). Lifetime prevalence of psychiatric disorders in Edmonton. *Acta Psychiatrica Scandinavia,* 77 (Suppl 338), 24-32.

Bloembergen, W. E., Port, F. K., Mauger, E. A. et al. (1996). Gender discrepancies in living related renal transplant donors and recipients. *Journal of the American Society of Nephrology,* 7, 1139–44.

Blom, M. & Jennissen, R. (2013). The involvement of different ethnic groups in various types of crime in the Netherlands. *European Journal of Crime and Policy Research*, 20, 51–72.

Blonigen, D.M., Carlson, S.R., Krueger, R.F. & Patrick, C.J. (2003). A twin study of self-reported psychopathic personality traits. *Personality and Individual Differences*, 35, 179-197.

Blum, R. W., Harmon, B., Harris, L., Bergeissen, L. & Resnic, M. D. (1992). American Indian-Alaskan youth health. *Journal of the American Medical Association*, 267, 1637–1644.

Blumenbach, J. F. (1776) *De generis humani varietate nativa liber.* Goettingen: Vandenhoek.

Blurton Jones, N. G., Marlowe, F. W., Hawkes, K., & O'Connell,

J. F. (2000). Paternal investment and hunter-gatherer divorce rates. In L. Cronk, N. Chagnon, & W. Irons (Eds) *Adaptation in human behaviour: An anthropological perspective* (pp 69–90). Chicago, IL: Aldine Transaction.

Boy, A. & Kulczycki, A. (2008). What we know about intimate partner violence in the Middle East and North Africa. *Violence Against Women*, 14, 53-68.

Boyer, C. B., Tschann, J. M. & Shafer, M. A. (1999). Predictors of risk for sexually transmitted diseases in ninth grade urban high school students. *Journal of Adolescent Research*, 14, 448–465.

Braun, B. L., Murray, D., Hannan, P., Sidney, S. & Le, C. (1996). Cocaine use and characteristics of young adult users from 1987 to 1992: The Cardia study. *American Journal of Public Health*, 86, 1736-1741.

Braun, B. L., Hannan, P., Wolfson, M., Jones-Webb, R. & Sidney, S. (2000). Occupational attainment, smoking, alcohol intake, and marijuana use: ethnic-gender differences in the Cardia study. *Addictive Behaviors*, 25, 399-414.

Breslau, J., Aguilar-Gaxiola, S., Kendler, K. S., Su, M., Williams, D. & Kessler, R. C. (2006). Specifying race-ethnic differences in risk for psychiatric disorder in a USA national sample. *Psychological Medicine*, 36, 57–68.

Breslau, J., Kendler, K. S., Su, M., Aguilar-Gaxiola, S. & Kessler, R. C. (2005). Lifetime risk and persistence of psychiatric disorders across ethnic groups in the United States. *Psychological Medicine, 35,* 317–327.

Broman, C. L. (1993). Race differences in marital well-being. *Journal of Marriage and the Family,* 55, 724–732.

Brooner, R. K, Greenfield, L., Schmidt, C,W & Bigelow, G. E. (1993). Antisocial personality disorder and HIV infection among intravenous drug users. *American Journal of Psychiatry, 150, 53-58.*

Brown, E. & Ferris, J. (2007). Social capital and philanthropy: An analysis of the impact of social capital on individual giving and volunteering. *Nonprofit and Voluntary Sector Quarterly,* 36, 85–99.

Bruner, H. G., Nelen, M, Breakfield, X. O., Ropers, H. H. & van Oost, B. A. (1993). Abnormal behaviour associated with a point mutation in the structural gene for monoamine oxidase A. *Science,* 262, 578-580.

Burke, B. M. (1997). Trends and compositional changes in fertility: California circa 1970-1990. *Population and Environment,* 19, 15–51.

Burrows, S. (2007). Sex-specific suicide mortality in the South African urban context: the role of age, sex, race, and geographical location. *Scandinavian Journal of Public Health,* 35, 133-139.

Burton, R. F. (1885). *The Book of the Thousand Nights and a Night.* London: The Burton Club.

Caetano, R., Field, G. A., Ramisetty-Mikler, S. & McGrath, C. (2005). The 5-year course of intimate partner violence among White, Black and Hispanic couples in the United States. *Journal of Interpersonal Violence,* 20, 1039–1057.

Cale, E. M. & Lilienfeld, S. O. (2006). Psychopathy factors and risk for aggressive behavior: a test of the threatened egotism hypothesis. *Law & Human Behavior,* 30, 51-74.

Campos-Outcalt, D., Bay, C., Dellapena, A. & Coya, M. K. (2003). Motor vehicle crash fatalities by race/ethnicity in Arizona, 1990-96. *Injury Prevention,* 9, 251-256.

Careal, M., Cleland, J., Deheneffe, J., Ferry, B., & Ingham, R. (1995). Sexual behavior in developing countries: Implications for HIV control. *AIDS, 9,* 1171–1175.

Carroll, E. (1977). Notes on the epidemiology of inhalants. *National Institute of Drug Research Monograph,* series 15, 2-11.

Carmen, J. A. & Roberts, M. A. (1934). *East African Medical Journal,* 11, 107-125. .

Carothers, J. C. (1953). *The African Mind in Health and Disease.* Geveva: World Health Organisation.

Cartwright, D. (2012). Borderline personality disorder: What do we know? Diagnosis, course, co-morbidity, and aetiology. *South African Journal of Psychology,* 38, 429-446.

Carver, J., Gaston, S., Altice, F., & Niccolai, L. (2014). Sexual risk behaviors among adolescents in Port-au-Prince, Haiti. *AIDS and Behavior,* 18, 1595-1603.

Cases, O., Seif, I., Grimsby, J. et al. (1995). Aggressive behaviour and altered amounts of brain serotonin and norepinephrine in mice lacking MAOA. *Science,* 268, 1763-1766.

Caspi, A., McClay, J., Moffitt, T. et al. (2002). Role of genotype in the cycle of violence in maltreated children. *Science,* 297 (5582), 851-854.

Catania, J.A., Coates, T.., Kegeles, S. et al. (1992). Condom use in multi-ethnic neighborhoods in San Francisco. *American Journal of Public Health, 82,* 284-287.

Catania, J. A., Coates, T. J., Golden, E. et al. (1994). Correlates of condom use among black, Hispanic and white heterosexuals in San Francisco. *AIDS Education and Prevention,* 6, 12-26.

Cavalli-Sforza, L. L., Menozzi, P. & Piazza, A. (1994). *The History and Geography of Human Genes.* Princeton, NJ: Princeton University Press.

Cazenaze, N.A. & Strauss, M.A. (1990). Race, class, network embeddedness and family violence. In M. A. Strauss & R. J. Gelles (Eds). *Physical Violence in American Families.* New Brunswick:

Transaction.

CDC (Centers for Disease Control and Prevention) (2008). Youth risk behavioral surveillance -United States, 2007. *MMWR, Surveillance Summaries,* 57, (SS-4).

CDC (Centers for Disease Control) (2012). *Today's HIV/AIDS epidemic.* Atlanta, GA: Centers for Disease Control.

CDC (Centers for Disease Control). (2012a). *Fact sheet: HIV and young men who have sex with men.* Atlanta, GA: Center for Disease Control.

Chabrol, H. Valls, M., van Leeuwen, N. & Bui, E. (2012). Callous-unemotional and borderline traits in nonclinical adolescents: personality profiles and relations to psychopathic behaviors. *Personality and Individual Differences,* 53, 969-973.

Chang, L., Morrissy, R. F. & Koplewicz, H. S. (1995). Prevalence of psychiatric symptoms and their relation to adjustment among Chinese-American youth. *Journal of the Academy of Child and Adolescent Psychiatry*, 34, 91–99.

Chen, C., Wong, J. & Lee, N. (1993). The Shatin community mental health survey in Hong Kong. *Archives of General Psychiatry*, 50, 125-133.

Child Youth and Family. (2006). *Department of child youth and family services annual report: The true measure of our success: Six years of achievement.* Wellington, New Zealand.

Children's Commissioner (2012). *Report on England on Child Sexual Exploitation.* London: Children's Commissioner.

Children's Defense Fund (1985). *Black and white children in America.* Washington, DC: Author.

Choi, K-H., Catania, J. A. & Dolcini, M. M. (1994). Extramarital sex and HIV risk behavior among US adults: results from the National AIDS Behavioral Survey. *American Journal of Public Health*, 84, 2003–2007.

Clarizio, H. E. (1997). Conduct disorder: developmental considerations. *Psychology in the Schools*, 34, 253-265.

Cleckley, H. (1941, 1976). *The mask of sanity*. St. Louis, MO: Mosby.

Coker, A. L., Richter, D. L. & Valois, R. F. (1994). Correlates and consequences of early initiation into sexual intercourse. *Journal of School Health*, 64, 372–377.

Coid, J. W., Yang, M., Ullrich, S., Roberts, A. & Hare, R. D. (2009). Prevalence and correlated of psychopathic traits in the household population of Great Britain. *International Journal of Law and Psychiatry*, 32, 65-73.

Coleman, A. (2008). *A Dictionary of Psychology* (3 ed.). Oxford University Press.

Colon, I. (1992). Race, belief in destiny and seat belt usage: a pilot study. *American Journal of Public Health*, 82, 875-877.

Compton, W. M., Conway, K. P., Stinson, F. S. et al. (2005). Prevalence, correlates, and comorbidity of DSM-IV antisocial personality syndromes and alcohol and specific drug use disorders in the United States: Results from the national epidemiologic survey on alcohol and related conditions. *Journal of Clinical Psychiatry*, 66, 677–685.

Compton, W. M., Cottler, L. B., Abdallah, A. B. et al. (2000). Substance dependence and other psychiatric disorders among drug dependent subjects: race and gender correlates. *The American Journal on Addictions*, 9, 113-125.

Compton, W. M., Helzer, J. E., Hwu, H. G. et al. (1991). New methods in cross-cultural psychiatry: psychiatric illness in Taiwan and the United States. *American Journal of Psychiatry*, 148, 1697-1704.

Conners, C. K. (1989). *Conners' rating scales manual*. Tonawanda, NY: Multi-Health Systems.

Cooke, D. J. & Michie, C. (2001). Refining the construct of psychopath: Towards a hierarchical model. *Psychological Assessment*, 13, 171-188.

Cooke, D. J., Mitchie, C., Hart, S. D. & Clark, D. (2005). Searching for the pan-cultural core of psychopathic personality disorder.

Personality and Individual Differences, 39, 283-295.

Coon, C. S., Garn, S. M. & Birdsell, J. B. (1950). *Races.* Springfield, Ill: Thomas.

Corr, P.J. (2010). The psychoticism-psychopathy continuum: a neuropsychological model of core deficits. *Personality and Individual Differences,* 48, 695-703.

Correctional Service of Canada (1990). A mental health profile of federally sentenced offenders. *Forum on Corrections Research*, 2, 7–8.

Cort, M. & Cort, D. (2008). Willingness to participate in organ donation among Black Seventh Day Adventist college students . *Journal of American College Health*, 56, 691 - 697.

Costenbader, V. K. & Markson, S. (1994). School suspension: a survey of current policies and practices. *NASSP Bulletin*, October, 103–107.

Cove, J. J. (1992). Aboriginal over-representation in prisons: what can be learned from Tasmania? *Australian and New Zealand Journal of Criminology,* 25, 156–168.

Cozzetto, D. & Larocque, B. (1996). Compulsive gambling in the Indian community: a North Dakota case study. *American Aboriginal Culture and Research Journal,* 20, 73-86.

Crawford, M. J., Rushwaya, T., Bajaj, P., Tyrer, P. & Yang, M. (2011). The prevalence of personality disorder among ethnic minorities: findings from a national household survey. *Personality and Mental Health*, 6, 175-183.

Cross, W.E. (1985). Black identity. In M.B. Spencer (Ed) *Beginnings: The social and affective development of black children.* Mahwah, N.J.: Lawrence Erlbaum.

Cuffe, S. P., McKeown, R. E., Addy, C. L. & Garrison, C. Z. (2005). Family and psychosocial risk factors in a longitudinal epidemiological study of adolescents. *Journal of the American Academy of Child and Adolescent Psychiatry,* 44, 121-129.

Cumberworth, D. (2004). *Australia.* In K.Malley-Morrison (Ed) *International Perspectives on Family Violence.* Mahwah, NJ: Lawrence

Erlbaum.

Currie, C.L. (2013). Illicit and prescription drug problems among Aboriginal adults in Canada. *Social Science & Medicine*, 88, 1-9.

Dabbs, J.M. (2000). *Heroes, Rogues, and Lovers: Testosterone and Behavior* . New York: McGraw-Hill.

Dabbs, J.M. & Morris, R. (1990). Testosterone, social class and psychopathic behavior in a sample of 4,462 men. *Psychological Science*, 1, 209-211.

Daderman, A.M. & Kristiansson, M. (2003). Degee of psychopathy – implications for treatment in male juvenile delinquents. *International Journal of Law & Psychiatry*, 26, 3-01-315.

Dahlstrom, W. G., Lachar, D. & Dahlstrom, L. E. (1986). *MMPI patterns of American minorities*. Minneapolis, MN: University of Minneapolis Press.

Darroch, J. E., Landry, D. J. & Oslak, S. (1999). Age differences between sexual partners. *Family Planning Perspectives*, 31, 160–167.

Davis, J. & Sorensen, J. R. (2013). Disproportionate minority confinement of juveniles: A national examination of black–white disparity in placements, 1997-2006. *Crime and Delinquency*, 59, 115-139.

Dawson, D. A. (1998). Beyond black, white and Hispanic: race, ethnic origin and drinking patterns in the United States. *Journal of Substance Abuse,* 10, 321-339.

Day, R.D. (1992). The transition to first intercourse among racially and culturally diverse youth. *Journal of Marriage and the Family,* 54, 749-762.

Dean, A. C., Altstein, L. L., Berman, M. E. et al. (2013). Secondary psychopathy, but not primary psychopathy, is associated with risky decision-making in noninstitutionized young adults. *Personality and Individual Differences*, 54, 272-277.

DeBerry, Kimberly (1991). *Modeling Ecological Competence in African American Transracial Adoptees.* Ph.D. Dissertation. University of Virgina, Charlottesville.

Debowska, A., Boduszek, D., Kola, S. & Hyland, P. (2014). A bifactor model of the Polish version of the Hare self-report psychopathy scale. *Personality and Individual Differences*, 69, 231-237.

DeCuyper, M., De Fruyt, F. & Buschman, J. (2008). A five-factor model perspective on psychopathy and comorbid Axis-II disorders in a forensic-psychiatric sample. *International Journal of Law and Psychiatry*, 31, 394-406.

Derevensky, J. L. (2008). Gambling behaviors and adolescent substance abuse disorders. In Y. Kaminer & O. G. Bukstein (Eds) *Adolescent Substance Abuse*. New York: Routledge.

Department of Indian Affairs and Northern Development Canada (2003). *Backgrounder: The residential school system*. Ottawa.

De Salas del Valle, H. (2009). *Afro-Cubans: Powerless majority in their own country*. University of Miami Institute for Cuban and Cuban-American Studies.

Diamond, B., Morris, R. G. & Barnes, J. C. (2012). Individual and group IQ predict inmate behavior. *Intelligence*, 40, 115-122.

Dion, R., Gotowiec, A. & Beiser, M. (1998). Depression and conduct disorder in native and non-native children. *Journal of the American Academy of Child and Adolescent Psychiatry*, 7, 736-742.

Doherty, O. & Matthews, G. (1988). Personality characteristics of opiate addicts. *Personality and Individual Differences*, 9,171-172.

Dolcini, M. M., Catania, J. A., Coates, T. J. et al., (1993). Demographic characteristics of heterosexuals with multiple partners: The national AIDS behavioral surveys. *Family Planning Perspectives*, 25, 208-214.

Donnan, S. (2001). Aboriginal sin is now the issue. *Financial Times*, 21 July.

Donoghue, E. (1992). Sociopsychological correlates of teen-age pregnancy in the United States Virgin Islands. *International Journal of Mental Health*, 21, 39-49.

Dookie, I. J. (2004). *Canada*. In K. Malley-Morrison (Ed) *International Perspectives on Family Violence*. Mahwah, NJ: Lawrence

Erlbaum.

Dowe, G., King, S.D., Smikle, M.F.et al. (1998). Prevalence of viral and bacterial sexually transmitted pathogens in Jamaican pregnant women. *West Indian Medical Journal*, 47, 23-25.

Dubanoski, R. A. & Snyder, K. (1980). Patterns of child abuse and neglect in Japanese and Samoan Americans. *Child Abuse and Neglect*, 4, 217-225.

Dunbar, R. (2010). Deacon's dilemma: The problem of pair-bonding in human evolution. In R. Dunbar, C. Gamble & J. Gowlett (Eds) *Social Brain, Distributed Mind*. Oxford: Oxford University Press.

Dunne, M.P., Martin, N.G, Statham, D.J. et al. (1997). Genetic and environmental contributions to variance in age at first sexual intercourse. *Psychological Science*, 8, 211-216.

DuPaul, G. J., Power, T. J., Anastopoulos, A. D., Reid, R., McGoey, K. E. & Ikeda, M. J. (1997). Teacher ratings of attention deficit hyperactivity disorder symptoms: factor structure and normative data. *Psychological Assessment*, 9, 436–444.

Durbin, M., DiClemente, R.J., Siegel, D. et al. (1993). Factors associated with multiple sexual partners among junior high school students. *Journal of Adolescent Health*, 14, 202-207.

Dutton, E. (2018). *J. Philippe Rushton: A Life History Perspective*. Oulu, Finland: Thomas Edward Press.

Dutton, E. & Lynn, R. (2014). Racial differences in psychopathic personality and cheating in sport. *Mankind Quarterly*, 55, 325-334.

East, P. (1998). Racial and ethnic differences in girls' sexual, marital and birth expectations. *Journal of Marriage and the Family*, 60, 150-162.

Eastwell, H.D. (1979). Petrol-inhalation in Aboriginal towns. (personal communication).

Eisenbrger, N.I., Way, B.M., Taylor, S.E. et al. (2007). Understanding genetic risk for aggression: clues from the brain's response to social exclusion. *Biological Psychiatry*, 61, 100-108.

Elion, V. H. & Megargee, E. I. (1975). Validity of the MMPI Pd scale among black males. *Journal of Consulting & Clinical Psychology*, 43, 166–172.

Ellickson, P. L. & Morton, S. C. (1999). Identifying adolescents at risk for hard drug use: racial/ethnic variations. *Journal of Adolescent Health*, 25, 382–395.

Elliott, D. S., Ageton, S. & Huizinga, D. (1980). *Self-reported delinquency estimates by sex, race, class and age*. Washington, DC: Behavioral Science Institute.

Ellis, L. (1990). Universal behavioral and demographic correlates of criminal behavior . In L. Ellis & H. Hoffman (Eds) *Crime in biological, social and moral contexts*. Westport, CT: Praeger.

Ellis, L., Hoskin, A., Hartley, R., Walsh, A., Widmayer, A. & Ratnasingam, M. (2014). General theory versus ENA theory: Comparing their predictive accuracy and scope. *International Journal of Offender Therapy and Comparative Criminology*, 59, doi:10.1177/0306624x14543263

Ellis, L. & Nyborg, H. (1992). Racial/ethnic variations in male testosterone levels: a probable contributor to group differences in health. *Steroids*, 57,72-75.

Ellison, P. T., Bribiescas, R. G., Bentley G. R. & Campbell, B. C. (2002). Population variation in age-related decline in male salivary testosterone. *Human Reproduction*, 17, 3251-3253.

Ellsberg, M. C., Pena, R., Herrera, A., Liljestrand, J., & Winkvist, A. (1999). Wife abuse among women of childbearing age in Nicaragua. *American Journal of Public Health*, 89, 241–244.

Epstein, J. N., March, J. S., Conners, C. K. & Jackson, D. L. (1998). Racial differences on the Conners Teacher Rating Scale. *Journal of Abnormal Child Psychology*, 26, 109–118.

Estrada-Martínez, L. M., Caldwell, C. H., Schulz, A. J. et al. (2013). Families, neighborhood sociodemographic factors, and violent behaviors among Latino, White, and Black adolescents. *Youth and Society*, 45, 221-242.

Eysenck, H. J. & Gudjonsson, G. H. (1989). *The causes and cures of criminality.* New York: Plenum.

Ezeh, A. C. (1997). Polygyny and reproductive behavior in sub-Saharan Africa. *Demography,* 34, 355-368.

Falkenbach, D. M., Howe, J. R. & Falki, M. (2013). Using self-esteem to disaggregate psychopathy, narcissism and aggression. *Personality and Individual Differences,* 54, 815-820.

Fals-Stewart, W. (2005). Substance abuse disorders. In J. E. Maddux & B.A. Winstead (Eds) *Psychopathology.* Mahwah, NJ: Lawrence Erlbaum.

Fan, M. S., Hong, J. H., Ng, .L. et al. (1995). Western influences on Chinese sexuality. *Journal of Sex Education and Therapy,* 21, 158-166.

Farrington, D. F. (1991). Psychopathic personality from childhood to adulthood. *The Psychologist,* 4, 389-394.

Feldmeyer, B. & Steffensmeier, D. (2013). Patterns and trends in elder homicide across race and ethnicity, 1985-2009. *Homicide Studies,* 17, 204-223.

Feng, H. & Cartledge, G. (1996). Social skill assessment of inner city Asian, African and European American students. *School Psychology Review,* 25, 228–239.

Fergusson, D., Horwood, J. & Swain-Campbell (2003). Ethnicity and criminal convictions. *Australian and New Zealand Journal of Criminology,* 36, 354-367.

Fernbrant, C., Emmelin, M., Essén, B., Östergren, P.-O., & Cantor-Graae, E. (2014). Intimate partner violence and poor mental health among Thai women residing in Sweden. *Global Health Action,* 7, 24991.

Ferreira da Silva, L. (1991). O direitode batar na mulher. *Analise Social,* 26, 385-397.

Fetters, W. B., Stowe, P. S. & Owings, J. A. (1984). *Quality of responses of high school students to questionnaire items.* Washington, DC: National Center for Educational Statistics.

Finer, L. B., Darroch, J. E. & Singh, S. (1999). Sexual partnership patterns as a behavioral risk factor for sexually transmitted diseases. *Family Planning Perspectives*, 31, 228–236.

Finkelhor, D. (1997). The homicides of children and youth. In G. K. Kantor & J. L. Jasinski (Eds) *Out of darkness*. London: Sage.

Fishburne, P.M., Abelson, H.I. & Cisin, I. (1980). *National Survey on Drug Abuse*. Washington, D.C. : U.S. Govt Printing Office.

Fisher, H. E. (1992). *Anatomy of love: A natural history of mating, marriage, and why we stray*. New York, NY: Fawcett Columbine.

Fisher, J., Trana, T. D., Biggs, B. et al. (2013). Intimate partner violence and perinatal common mental disorders among women in rural Vietnam. *International Health*, 5, 29-37.

Flanaghan, T. J. (1980). Correlates of institutional misconduct among state prisoners: A research note. *Criminology*, 21, 41-65.

Flaskerud, J. H. & Hu, L.-T. (1992). Relationship of ethnicity to psychiatric diagnosis. *Journal of Mental & Nervous Disease*, 10, 296-303.

Flint, A. J., Yamada, E. G. & Novotny, T. E. (1998). Black-white differences in cigarette smoking uptake: progression from adolescent experimentation to regular use. *Preventative Medicine*, 27, 358–364.

Fluke, J.D., Chabot, M. & Fallon, B. (2013). Placement decisions and disparities among Aboriginal groups. *Child Abuse and Neglect*, 37, 47-60.

Fluke, J. D., Yuan, Y.-Y., Hedderson, J. & Curtis, P. A. (2003). Disproportionate representation of race and ethnicity in child maltreatment. *Children and Youth Services Review*,.25, 359-393.

Flynn, J. R. (1991). *Asian Americans: achievement beyond IQ*. Hillsdale, NJ: Erlbaum.

Foldes, H. J., Duehr, E. E. & Ones, D. S. (2008). Group differences in personality: meta-analyses comparing five U.S. racial groups. *Personnel Psychology*, 61, 579-616.

Fontova, H. (2005). *Fidel: Hollywood's Favorite Tyrant*. Washington,

D.C: Regnery Publishing.

Foran,T. (1995). A descriptive comparison of demographic and family characteristics of the Canadian and offender populations. *Forum on Corrections Research Corrections Services Canada, 7,* 2-16.

Ford, K., Zelnik, M. & Kantner, J. F. (1981). Sexual behavior and contraceptive use among socioeconomic groups of young women in the United States. *Journal of Biosocial Science,* 13, 31-45.

Forero, R., Bauman, A., Chen, J. X. & Flaherty, B. (1999). Substance use and socio-demographic factors among aboriginal and Torres Strait Islander school students in New South Wales. *Australian and New Zealand Journal of Public Health,*23, 295-300.

Forrest, D. & Wardle, H. (2011). Gambling in Asian communities in Great Britain. *Asian Journal of Gambling Issues and Public Health,* 2, 2-16.

Forrest, J. D. & Singh, S. (1990). The sexual and reproductive behavior of American women, 1982-1988. *Family Planning Perspectives,* 22, 206–214.

Forste, R. & Tanfer, K. (1996). Sexual exclusivity among dating, cohabiting and married women. *Journal of Marriage and the Family,* 58, 33-47.

Fossian, P., Ledoux,Y., Valente, F. et al. (1998). Psychiatric disorders and social characteristics among second generation Moroccan migrants in Belgium. *European Psychiatry,* 17, 443-450.

Fourie, R. (2004). South Africa. In K. Malley-Morrison (Ed) *International Perspectives on Family Violence.* Mahwah, NJ: Lawrence Erlbaum.

Fraser, S. L., Muckle, G., Abdous, B. B. et al. (2012). Effects of binge drinking on infant growth and development in an Inuit sample. *Alcohol,* 45, 277-283.

French, B.H. & Neville, H.A. (2013). Sexual coercion among black and white teenagers. *Counseling Psychology,* 41, 1186-1212.

Friedman, J. J., McFarlane, C. P. & Morris, L. (1997). *Jamaica Reproductive Health Survey 1997.* Kingston: National Family Planning

Board.

Friedemann-Sanchez, G (2012). Intimate partner violence in Columbia: who is at risk? *Social Forces*, 91, 663-688.

Fryer, P. (1984). *Staying Power*. London: Pluto Press.

Furstenberg, F. F., Morgan, S. P., Moore, K. A. & Peterson, .L. (1987). Race differences in the timing of adolescent intercourse. *American Sociological Review*, 52, 511-518.

Gada, M. (1987). A study of prevalence and pattern of attention deficit disorder with hyperactivity in primary school children. *Indian Journal of Psychiatry*, 29, 113-118.

Galton, F. (1908) *Memories of My Life*. London: Methuen.

Gao, Y., Raine, A. Venables, P. H., Dawson, M. E. & Mednick, S. A. (2010). Association of poor childhood fear conditioning and adult crime. *American Journal of Psychiatry*, 167, 56-60.

Gardner, R. (1994). *Mortality*. In N. W. Zane, D. T. Takeuchi & K. N. Young (Eds). *Confronting Critical Health Issues of Asian & Pacific Islander Americans*. Thousand Oaks, CA: Sage.

Gazmararian, J. A., Adams, M. M. & Saltzman, L. E. (1995). The relationship between pregnancy intendedness and physical violence in mothers of newborns. *Obstetrics and Gynecology*, 85, 1031-1037.

Giammarco, E. A., Atkinson, B., Baughman, H. M. Veselka, L. & Vernon, P. A. (2013). The relation between antisocial personality and the perceived ability to deceive. *Personality and Individual Differences*, 54, 246-250.

Gielen, U. P., Reid, C. & Avellani, J. (1989). Perceptions of parental behavior and the development of moral reasoning in Jamaican students. In L. L. Adler (Ed) *Cross cultural research in human development*. Westport, CT: Praeger.

Gillborn, D. & Gipps, C. (1996). *Recent research on the achievements of ethnic minority pupils*. London: Office for Standards in Education.

Gladden, P. R., Figueredo, A. J. & Jacobs, J. (2008). Life history strategy, psychopathic attitudes, personality, and general intelligence.

Personality and Individual Differences, 46, 270-275.

Gladue, B.A. (1991). Aggressive behavioral characteristics, hormones, and sexual orientation in men and women. *Aggressive Behavior*, 17, 313-326.

Glenn, A. L., Johnson, A. K. & Raine, A. (2013). Antisocial personality disorder: a current review. *Current Psychiatry Reports*, 15, 427-435.

Goetting, A. (1989). Patterns of marital homicide: a comparison of husbands and wives. *Journal of Comparative Family Studies*, 20, 341–354.

Gold, M. (1966). Undetected criminal behavior . *Journal of Research in Crime and Delinquency*, 3, 27-46.

Goldberg, L. R., Sweeney, D., Merenda, P. & Hughes, J. E. (1998). Demographic variables and personality: the effects of gender, age, education and ethnic/racial status on self-descriptions of personality attributes. *Personality and Individual Differences*, 24, 393-403.

Goldenberg, D. M. (2003). *The Curse of Ham. Race and Slavery in Early Judaism, Christianity, and Islam*. Princeton, NJ: Princeton University Press.

Goldman, V. J., Cooke, A. & Danistrom, G. (1995). Black-white differences among college students: a comparison of MMPI and MMPI-2 norms. *Assessment*, 2, 293-299.

Golub, A. & Johnson, B. D. (2005). The new heroine users among Manhattan arrestees: variations by race/ethnicity and mode of consumption. *Journal of Psychoactive Drugs*, 37, 51-61.

Gomez, S. L., Le, G. M., Clarke, C. A. et al. (2003). Cancer incidence patterns in the US and in Kangwha, South Korea. *Cancer Causes and Control*, 14, 167-174.

Gonzales, R. A., Kallis, C., Ulrich, S. et al. (2014). The protective role of higher intellectual functioning on violence in the household population of Great Britain. *Personality and Individual Differences*, 61, 82-85.

Goodman, R. & Richards, H. (1995). Child and adolescent

psychiatric presentations of second generation Afro-Caribbeans in Britain. *British Journal of Psychiatry*, 167, 362–369.

Gordon, R., Piana, L. D. & Keleher, T. (2000). *Facing the consequences: an examination of racial discrimination in US public schools*. Oakland, CA: Applied Research Center.

Gordon, V., Donnelly, P. D. & Williams, D. J. (2014). Relationship between ADHD symptoms and anti-social behavior in a sample of older youths in adult Scottosh prisons. *Personality and Individual Differences*, 58, 116-121.

Grant, J.E. & Potenza, M.N. (2012). *The Oxford Handbook of Impulse Control Disorders*. Oxford: Oxford University Press.

Gray, D., Morfitt, B., Ryan, K. & Williams, S. (1997). The use of tobacco, alcohol and other drugs by Aboriginal young people in Albany, Western Australia. *Australian and New Zealand Journal of Public Health*, 21,71-76.

Gray, M. C., Hunter, B. & Schwab, R. G. (2000). Trends in indigenous educational participation and attainment, 1986–1996. *Australian Journal of Education*, 44, 101–117.

Green, H. B. (1964). Socialization values in Negro and East Indian subcultures of Trinidad. *Journal of Social Psychology*, 64, 1-20.

Green, H. B. (1972). Temporal attitudes in four Negro subcultures. In J. T. Fraser, F. C. Haber & G. H. Miller (Eds) *The Study of Time*. New York: Springer-Verlag.

Gregory, S., Ffytch, D., Simmons, A. et al. (2012). The psychopathic brain: psychopathy matters. *Archives of General Psychiatry*, 69, 962-972.

Grinstead, O.A., Faigeles, B., Binson, D. & Eversley, R. (1993). Sexual risk taking for human immunodeficiency virus infection among women in high-risk cities. *Family Planning Perspectives*, 25, 252-256.

Grove, W.M., Eckert, E.D., Frick et al. (1990). Heritability of substance abuse and antisocial behavior: a study of monozygotic twins reared apart. *Biological Psychiatry*, 27, 1293-1304.

Gruber, E., Di Clemente, R. J. & Anderson, M. M. (1996). Risk-taking behavior among Native American adolescents in Minnesota public schools: comparisons with black and white adolescents. *Ethnicity and Health*, 1, 261–267.

Giluk, T.L. & Postlethwaite, B.E. (2015).Big five personality and acidic dishonesty: a meta-analytic review. *Personality and Individual Differences*, 72, 59-67.

Gupta, N. (2000). Sexual initiation and contraceptive use among adolescent women in Northeast Brazil. *Studies in Family Planning*, 31, 228-238.

Gurege, O., Lasebikan, V.O., Kola, L. & Makanjuola, V.A. (2006). Lifetime and 12-month prevalence of mental disorders in the Nigerian Survey of Mental Health and Well-being. *British Journal of Psychiatry*, 188, 465-471.

Guttmacher, A. (1994). *Sex and America's Teenagers*. New York: Alan Guttmacher Institute.

Guttmacher, S., Lieberman, L., Ward, D. et al. (1997). Condom availability in New York city public high schools. American *Journal of Public Health, 87, 1427-1433.*

Guze, S. B. (1976). *Criminality and psychiatric disorders*. New York: Oxford University Press.

Halpern, C. T., Udry, J. R. & Suchindran, C. (1997). Testosterone predicts initiation of coitus in adolescent females. *Psychosomatic Medicine*, 59, 161-171.

Halpern, C. T., Udry, J. R. & Suchindran, C. (1998). Monthly measures of salivary testosterone predict sexual activity in adolescent males. *Archives of Sexual Behavior*, 27, 445-465.

Halpern, R. (2004). Solving the labor problem: race, work, and the state of the sugar industries of Louisiana and Natal, 1870–1919. *Journal of Southern African Studies*, 30, 19–40.

Hampton, R. L., Gelles, R. J. & Harrop, J. W. (1989). Is violence in black families increasing? *Journal of Marriage and the Family*, 51, 969–980.

Harden, T. (2014). Race differnces in crime. *Sunday Times*, 17 August, p 23.

Hare, R. D. (1983). Diagnosis of psychopathic personality disorder in two prison populations. *American Journal of Psychiatry*, 140, 887–890.

Hare, R. D. (1991). *Manual for the Hare Psychopathy Checklist -revised*. Toronto: Multi-Health Systems.

Hare, R. D. (1994). *Without conscience*. London: Warner.

Hare, R. D. (2003). *The Hare Psychopathy Checklist—Revised (PCL-R) manual* (2nd ed.). Toronto: Multi-Health Systems.

Hare, R. D. (2006). Psychopathy: a clinical and forensic overview. *Psychiatric Clinics of North America*, 29, 709-724.

Harpur, T. J., Hare, R. D. & Hakstian, R. (1989). Two-factor conceptualization of psychopathy: Construct validity and assessment implications. *Psychological Assessment: A Journal of Consulting and Clinical Psychology*, 1, 6-17.

Harpur, T. J., Hart, S. D. & Hare, R. D. (1994). The personality of the psychopath. In P. T. Costa & T. A. Widiger (Eds) *Personality disorders and the five factor model of personality*. Washington, DC: American Psychological Association.

Harrendorf, M., Heiskanen, M. & Malby, S. (2010). *International statistics on Crime and Justice*. Helsinki: European Institute for Crime Prevention and Control.

Harris, A. R. & Stokes, R. (1978). Race, self-evaluation and the protestant ethic. *Social Problems*, 26, 71-85.

Harris, J. A., Rushton, J. P., Hampson, E. & Jackson, D. N. (1996). Salivary testosterone and self-report aggression and pro-social personality characteristics in men and one. *Aggressive Behavior*, 22, 321-331.

Hart, S. D., Forth, A. E. & Hare, R. D. (1990). Performance of criminal psychopaths on selected neuropsychological tests. *Journal of Abnormal Psychology*, 99, 374-379.

Hathaway, S. R. & McKinley, J. C. (1940). A multiphasic personality schedule: construction of the schedule. *Journal of Psychology*, 10, 249-254.

Hathaway, S. R. & McKinley, J. C. (1989). *MMPI-2 manual*. Minneapolis, MN: University of Minnesota Press.

Hawaii (1987). *Crime in Hawaii*. Honolulu, HI: Office of the Attorney General.

Hawaii Dept. of Health (1990). *Statistical Report*. Honolulu, HI.

Hawkins, D. F. (2003). *Violent Crime: Assessing Race and Ethnic Differences*. Cambridge: Cambridge University Press.

Hayase, Y. & Liaw, K-L. (1997). Factors on polygamy in sub-Saharan Africa. *Developing Economies*, 35, 293-327. .

Health Canada (1997). *Family violence in Aboriginal communities*. Ottawa: Ntional Claringhouse on Family Violence

Henshaw, S. K. (1998). Unintended pregnancy in the United States. *Family Planning Perspectives*, 30, 24-29.

Herman-Giddens, M.E., Slora, E.J. & Wasserman, R.C. (1987). Secondary characteristics and menses in young girls. *Pediatrics*, 89, 505-512.

Herrnstein, R. J. & Murray, C. (1994). *The Bell Curve*. New York: Free Press.

Himle, J. A., Baser, R. E., Taylor, R. J., Campbell, R. D. & Jackson, J. S. (2009). Anxiety disorders among African Americans, blacks of Caribbean descent, and non-Hispanic whites in the United States. *Journal of Anxiety Disorders*, *23*, 578-590.

Hindelang, M. J. Hirschi, T. & Weis, J. G. (1981). *Measuring Delinquency*. Beverly Hills, CA: Sage.

Hofferth, S. L. & Hayes, C. D. (1987). *Risking the future: adolescent sexuality, pregnancy and childbearing*. Washington, DC: National Academy Press.

Hofferth, S. L., Kahn, J. R. & Baldwin, W. (1987). Premarital sexual activity among US teenage women over the last three decades. *Family*

Planning Perspectives, 19, 46-53.

Hogg, R. S. (1995). Aboriginal and non-Aboriginal mortality in rural Australia. *Human Organization,* 54, 214-221.

Hollander, D. (1996). Nonmarital childbearing in the United States: a government report. *Family Planning Perspectives*, 28, 29-41.

Holmes, S. A. (1999). *Survey details problems of minorities and credit. New York Times*, 22 September, p.2.

Home Office (1998). *Statistics on race and the criminal justice system.* London: HMSO.

Hoskin, A. W. & Ellis, L. (2015). Fetal testosterone and criminality: test of evolutionary neuroandrogenic theory. *Criminology*, 53, 54-73.

Hou, S-I. & Basen-Engquist, K. (1997). Human immunodeficiency virus risk behavior among white and Asian/Pacific Islander high school students in the United States. *Journal of Adolescent Health,* 20, 68-74.

Houston, L. N. (1981). Romanticism and eroticism among black and white students. *Adolescence*, 16, 263–272.

Howell, N. (1979). *Demography of the Dobe !Kung.* New York, NY: Walter de Gruyter.

Huang B., Grant, B.F., Dawson, D.A. et al. (2006). Race-ethnicity and the prevalence and co-occurrence of Diagnostic and Statistical Manual of Mental Disorders, Fourth Edition, alcohol and drug use disorders and Axis I and II disorders: United States, 2001 to 2002. *Comprehensive Psychiatry*, 47, 252-257.

Huizinger, D. & Elliott, D. S. (1984). *Self-reported measures of delinquency and crime.* Boulder, CO: Behavioral Research Institute.

Human Rights and Equal Opportunity Commission (2005). *Indigenous young people with cognitive disabilities and juvenile justice systems.* Sydney: Human Rights and Equal Opportunity Commission.

Human Sciences Research Council (2009). *South African HIV Prevalence, Incidence, Behaviour and Communication Survey, 2008.* Pretoria: Human Sciences Research Council.

Hunt, H. (1981). Alcoholism among Aboriginal people. *Medical Journal of Australia,* 1, 1-3.

Hunter, E. M., Hall, W. D. & Spargo, R. M. (1992). Patterns of alcohol consumption

in the Kimberley Aboriginal population. *Medical Journal of Australia,* 156, 764–768.

Hunter, W. W., Campbell, B. J. & Stewart, J. R. (1986). Seat belts pay off: the evaluation of a community-wide incentive program. *Journal of Safety Research,* 17, 23–31.

Hur, Y-M. & Bouchard, T. J. (1997). The genetic influence between impulsivity and sensation-seeking. *Behavior Genetics,* 27, 455-463.

Hwu, H. G., Yeh, E. K. & Chang, L. Y. (1989). Prevalence of psychiatric disorders in Taiwan. *Acta Psychiatrica Scandinavia,* 79, 136-147.

IBGE (Brazil government institute for statistics). (2008). *Statistical Report.* Sao Paulo.

India, B., Graham, V., Robertson, J. et al. (2011). High prevalence of cannabis use, mental health impacts, and potential intervention strategies: data from the Cape York cannabis project. *Drug & Alcohol Review,* 30, 13.

India, B. & Clough, A. R. (2012). Cannabis use in Cape York indigenous communities: high prevalence, mental health impacts and the desire to quit. *Drug and Alcohol Review,* 31, 580-584.

ING US (2012). *Retirement Revealed.* Washington, D.C.: Retirement Research Institute.

Ipsos Mori (2009). *British Survey of National Lottery and Gambling, 2008-9.* London: Ipsos Mori.

Jamaica Ministry of Health (2000). *National HIV/STD Prevntion and control programme annual report.* Kingston: Ministry of Health.

Jankowiak, W. R., & Fischer, E. F. (1992). A cross-cultural perspective on romantic love. *Ethnology,* 31, 149–155.

Jasinski, J. L. & Williams, L. M. (1998). *Partner Violence.* Thousand

Oaks, CA: Sage.

Japanese MMPI (1993). *New Japanese MMPI manual.* Kyoto: Sankyou-Bou.

Jencks, C. (1992). *Rethinking social policy: race, poverty and the underclass.* Cambridge, MA: Harvard University Press.

Jayakody, A. A., Viner, R. M., Haines, M. M. et al. (2006). Illicit and traditional drug use among ethnic minority adolescents in East London. *Public Health, 120,* 329-338.

Johansson, A., Grant, J., Kim, S., Odlaug, B. & Götestam, K. G. (2009). Risk factors for problematic gambling: a critical literature review. *Journal of Gambling Studies,* 25, 67-92.

Johnson, A.M., Wadsworth, J., Wellings, K. & Field, J. (1994). *Sexual Attitudes and Lifestyles.* Oxford: Blackwell.

Johnston, H.H. (1930). *A History of the Colonization of Africa by Alien Races.* Cambridge: Cambridge University Press.

Jones, J. R. & Schaubroeck, J. (2004). Mediators of the relationship between race and organizational citizenship behavior. *Journal of Managerial Issues, 16,* 505–527.

Joun,Y., Coonan, P. R. & LeGrande, M. E. (1997). Attitudes of Korean-American s in and around New York City toward organ transplantation. *Transplant Proceedings,* 29, 3751- 3752.

Junger, M. & Polder, W. (1993). Religiosity, religious climate and delinquency among ethnic groups in the Netherlands. *British Journal of Criminology,* 33, 416–435.

K. I. (1960). Kenya. *Encyclopedia Britannica.* Chicago: Encyclopedia Britannica.

Kaaya, S. F., Flisher, A. J., Mbwambo, J. K., Schaalma, H., Aarø, L. E. & Klepp, K.(2002). Review Article: A review of studies of sexual behaviour of school students in sub-Saharan Africa. *Scandinavian Journal of Public Health,* 30, 148-160.

Kahn, J. R., Rindfuss, R. R. & Guilkey, D. K. (1990). Adolescent contraceptive method choices. *Demography,* 27, 323-335.

Kalmuss, D. (1992). Adoption and black teenagers: the viability of a pregnancy resolution. *Journal of Marriage and the Family*, 54, 485–496.

Kang, S. V., Magura, S. & Shapiro, J. L. (1994). Correlates of cocaine/crack use among inner city incarcerated adolescents. *American Journal of Drug & Alcohol Abuse*, 20, 413-429.

Kapadia, F. (2013). Social support network characteristics. *Aids and Behavior*, 17, 1819-1828.

Kapiga, S.H., Lwihula, G.K., Shao, J.F. et al. (1993). Predictors of high risk sexual behavior and condo use among women in Dar es Salaam, Tanzania. *VIIIth Congress on AIDS in Africa, Marrakesh.*

Kastner, R.M. & Sellbom, M. (2012). Hyper-sexuality in college students: The role of psychopathy. *Personality & Individual Differences*, 53, 644-649.

Kenney, J. W., Reinholtz, C. & Angelini, P. (1997). Ethnic differences in childhood and adolescent sexual abuse and teenage pregnancy. *Journal of Adolescent Health*, 26, 210–263.

Kerr, W. C., Patterson, D. & Greenfield, T. K. (2009). Differences in the measured alcohol content of drinks between black, white and Hispanic men and women in a US national sample. *Addiction*, 104, 1503-1511.

Kessler, R. C., McGonagle, K. A., Zao, S. et al. (1994). Lifetime and 12-month prevalence of DSM-III-R psychiatric disorders in the United States. *Archives of General Psychiatry*, 51, 8-19.

Khurana, A., Romer, D., Betancourt, L., Brodsky, N. L., Gianetta, J. M. & Hurt, H. (2012). Early adolescent sexual debut: the mediating role of working memory ability, sensation seeking and impulsivity. *Developmental Psychology*, 48, 1416-1428.

Kiehl, K. A. & Sinnott-Armstrong, W. P. (2013). *Handbook on Psychopathy and Law.* Oxford: Oxford University Press.

Kishor, S. & Johnson, K. (2004). *Profiling Domestic Violence: A Multi-country Study.* Calverton, MD: ORC Macra.

Klomegah, R. (1999). Socio-demographic characteristics of

contraceptive users in Ghana. *Marriage and Family Review*, 29, 21-34.

Khawaja, M. & Barazi, R. (2005). Prevalence of wife beating in Jordanian refugee camps: Reports by men and women. *Journal of Epidemiology & Community Health*, 59, 840–841.

Koenig, M. A., Stephenson, R., Ahmed, S., Jejeebhoy, S. J. & Campbell, J. (2006). Individual and contextual determinants of domestic violence in North India. *American Journal of Public Health*, 96, 132–138.

Kohlberg, L. (1976). *Moral stages and moralization*. In T. Likona (Ed) *Moral development and behavior*. New York: Holt, Rinehart and Winston.

Kolsrud, K. (2006). To av tre ranere har innvandrerbakgrunn (two out of three robbers are immigrants). *Aftenpost* (Oslo), 28 Oct.

Kong, G., Tsai, J., Pilver, C.E. et al. (2010). Differences in gambling problem severity and gambling and health/functioning characteristics among Asian-American and Caucasian high-school students. *Psychiatry Research*, 172, 1071-1078.

Kraepelin, E. (1904). *Psychiatrie: Ein Lehrbuch*. Leipzig: Barth.

Kraft, J.M. & Coverdill, J.E. (1994). Employment and the use of birth control by sexually active single Hispanic, black and white women. *Demography*, 31, 593-602.

Kunitz, S. J., Gabriel, K. R. & Levy, J. E. (1999). Risk factors for conduct disorder among Navajo Indian men and women. *Social Psychiatry and Psychiatric Epidemiology*, 34, 180–189.

Kubricht W.S., Williams, B.J., Whatley, T., Pinckard, P. & Eastham, J.A. (1999). Serum testosterone levels in African-American and white men undergoing prostate biopsy. *Urology*, 54,1035-1038.

Kyriacou, D.N., Anglin, D. & Taliaferro, E. (1999). Risk factors for injury to women from domestic violence. *New England Journal of Medicine*, 341,1892-1898.

Lange, J. (1931). *Crime as Destiny*. London: Allen & Lane.

Langsdorf, R., Anderson, R. P., Waechter, D., Madrigal, J. F. & Juarez, L. J. (1979). *Psychology in the Schools*, 16, 293–298.

Lauderdale, M., Valiunas, A. & Anderson, R. (1980). Race, ethnicity and child maltreatment: an empirical analysis. *Child Abuse and Neglect*, 4, 163–169.

Lawrence, J. S., Marx, B. P., Scott, C.P. et al. (1995). Cross-cultural coparison of US and Nigerian adolescents' HIV-related knowledge, attitudes, and risk reduction interventions. *Aids Care*, 7, 499-461.

Leal, M. do Carmo (2006). Desigualdades raciais, sociodemographicias e na assistencia ao pre-natal e do parto, 1999-2001. *Saude da populacao negra no estado de Sao Paulo, Supplemento 6 do Boletim Epidemiologico Paulista*, 3, 36-45.

Lee, C.K., Kwak, Y.S, & Yamamoto, J. (1990). Psychiatric epidemiology in Korea. *Journal of Nervous and Mental Disease*, 178, 242-252.

Lee, J. (2013). Disadvantaged Aboriginal background now a factor in court rulings. *The Age*, October 3.

Lee, R. & Chambers, G. (2007). Monoamine oxidase, addiction, and the "warrior gene" hypthosis. *New Zealand Medical Journal*, 120, U2441.

Leigh, B. C., Temple, M. T. & Trocki, K. F. (1993). The sexual behavior of US adults: results from a national survey. *American Journal of Public Health*, 83, 1400–1408.

Leigh, B. C., Morrison, D. M., Trocki, K. F. & Temple, M. T. (1994). Sexual behavior of American adolescents: results from a US national survey. *Journal of Adolescent Health*, 15, 117-125.

Leistico, A. M. R., Salekin, R. T., DeCoster, J. & Rogers, R. (2008). A large-scale meta-analysis relating the hare measures of psychopathy to psychopathic conduct. *Law & Human Behavior*, 32, 28-45.

Lesieur, H. R., Blume, S. B . & Zoppa, R. M. (1986). Alcoholism, drug abuse and gambling. *Alcoholism: Clinical and Experimental Research*, 10, 33-38.

Leslie, L. M., Snyder, M. & Glomb, T. M. (2013). Who gives?

Multilevel effects of gender and ethnicity on workplace charitable giving. *Journal of Applied Psychology,* 98, 49–62.

Lester, D. (1989). Personal violence (suicide and homicide) in South Africa. *Acta Psychiatrica Scandinavia,* 79, 235-237.

Lester, D. (1998).*Suicide in African Americans.* New York: Nova Science, Lewis, B. (1990). *Race and Slavery in the Middle East.* New York: Oxford University Press.

Lichter, D. T., McLaughlin, D. K., Kephart, G. & Landry, D. J. (1992). Race and retreat from marriage: a shortage of marriageable men? *American Sociological Review,* 57, 781–799.

Liles, S., Usita, P. & Irvin, V. L. (2012). Prevalence and correlates of intimate partner violence among young, middle and older women of Korean descent in California. *Journal of Family Violence,* 27, 801-812.

Lim, S., Lambie, I. & Cooper, E. (2012). New Zealand youth that sexually offend: improving outcomes for Maori Rangatahi and their Wha–nau. *Sexual Abuse,* 12, 1-20.

Linnaeus, C. (1758*) Systema naturae sistens animale Sveciae regni.* Holmae: Salvius.

Linos, N. (2013). Influence of community social norms on spousal violence: Nigeria *American Journal of Public Heath,* 103, 148-155.

Lloyd, T., Hounsome, L., Mehay, A. et al. (2015). Lifetime risk of being diagnosed with, or dying from, prostate cancer by major ethnic group in England 2008–2010. *BMC Medicine,* 13, 171-191.

Loeber, R. (1990). Development and risk factors of juvenile psychopathic behavior and delinquency. *Clinical Psychology Review,* 10, 1-41.

Loo, S. K. & Rapport, M. D. (1998). Ethnic variations in children's problem behaviors: a cross-sectional developmental study of Hawaiian school children. *Journal of Child Psychology and Psychiatry,* 39, 567–575.

Lopes, F. (2006). Vamos fazar um teste: quai e a sua cor? *Saude da populacao negra no estado de Sao Paulo, Supplemento 6 do Boletim*

Epidemiologico Paulista, 3, 84-88.

Lorager, A. W., Janca, A. & Satorius, N. (1997). *Assessment and Diagnosis of Personality Disorders.* Cambridge: University Press.

Luk, S. L. & Leung, P. W. (1989). Conners' teacher's rating scale - a validity study in Hong Kong. *Journal of Child Psychology and Psychiatry,* 30, 785–793.

Lutalo, T., Kidugavu, M., Wawer, .J. et al. (2000). Trends and determinants of contraceptive use in Rakai district, Uganda, 1995-98. *Studies in Family Planning,* 31, 217-226.

Luzzo, D. A. (1994). An analysis of gender and ethnic differences in college students' commitment to work. *Journal of Employment Counseling,* 31, 38–46.

Lykken, D. T. (1995). *The psychopathic personalities.* Hillsdale, NJ: Lawrence Erlbaum.

Lynam, D., Moffit, T. & Stoutamer-Loeber, M. (1993). Explaining the relation between IQ and delinquency: class, race, test-motivation, school failure, or self-control? *Journal of Abnormal Psychology,* 102, 187-196.

Lynn, R. (1991). The evolution of race differences in intelligence. *Mankind Quarterly,* 32, 255–296.

Lynn, R. (1997). Geographical variation in intelligence. In H. Nyborg (Ed) *The scientific study of human nature.* Oxford: Pergamon.

Lynn, R. (2000). Race differences in sexual behavior and their demographic implications. *Population & Environment,* 22, 73-82.

Lynn, R. (2002). Racial and ethnic differences in psychopathic personality. *Personality & Individual Differences,* 32, 273-316.

Lynn, R. (2006). *Race Differences in Intelligence: An Evolutionary Analysis.* Second edition. Augusta, GA: Washington Summit Publishers. Second edition, 2015.

Lynn, R. (2008). *The Global Bell Curve.* Augusta, GA: Washington Summit Publishers.

Lynn, R. (2009). Race differences in school suspensions and exclusions in the United States. *Mank6nd Quarterly*, 50, 95-105.

Lynn, R. (2013). An examination of Rushton's theory of differences in penis length and circumference and r-*K* life history theory in 113 populations. *Personality and Individual Differences*, 55, 261-266.

Lynn, R. & Cheng, H. (2016). Ethnic and differences in conduct disorders, anti-social and prosocial behaviour among 11 year olds in the United Kingdom. *Mankind Quarterly*, 56, 572-579.

Lynn, R. & Cheng, H. (2016). Ethnic differences in conduct disorders, anti-social and pro-social behaviour among 11 year olds in the United Kingdom. *Mankind Quarterly*, 56, 572-59.

Lyons, M., Healy, N. & Bruno, D. (2013). It takes one to know one: relationship between lie detection and psychopathy. *Personality and Individual Differences*, 55,676-679.

MacGill, H. G. (1938). The Oriental delinquent in the Vancouver juvenile court. *Sociology and Social Research*, 12, 428-438.

Macy, M., Cunningham, R., Resnicow, K. & Freed, G. (2014). Disparities in age appropriate child passenger restraint used among children aged 1 to 12. *Pediatrics*, 133: 262-272.

Mancall, P. C., Robertson, P. & Huriwai, T. (2000). Maori and alcohol: a reconsidered history. *Australian and New Zealand Journal of Psychiatry*, 40, 129–134.

MAOA (2014). http://www.ensembl.org/Homo_sapiens/Variation/Population?db=core;r=X:43590536-43591536;v=rs6323;vdb=variation;vf=31754. Accessed February 15, 2018.

Macy, M.L., Cunningham, R. M., Resnicow, K. & Freed, G. L. (2014). Disparities in age-appropriate child passenger restraint use among children aged 1 to 12 years. *Pediatrics,*133, 262-271.

Malzberg, B (1944). Mental disease among American negroes. In O.Klineberg (Ed) *Characteristics of the American Negro*. New York: Harper.

Manning, W. D., & Cohen, J. A. (2012). Premarital cohabitation and marital dissolution: An examination of recent marriages. *Journal of Marriage and Family*, 74, 377–387.

Marie, D., Fergusson, D. M. & Boden, J. M. (2014). Childhood socio-economic status and ethnic disparities in psychosocial outcomes in New Zealand. *Australian and New Zealand Journal of Psychiatry,* 48, 679-680.

Marcin, J. P., Pretzlaff, R. K., Whittaker, .L. & Kon, A. A. (2003). Evaluation of race and ethnicity on alcohol and drug testing of adolescents admitted with trauma. *Academic Emergency Medicine,* 10, 1253-1259.

Marks, P. A., Seeman, W. & Haller, D. L. (1974). *The actuarial use of the MMPI with adolescents and adults.* Baltimore: Williams and Wilkins.

Marsh, A. A., Stoycosa, S. A., Brethel-Haurwitza, K. M. et al., (2014). Neural and cognitive characteristics of extraordinary altruists. *PNAS,* 111, 15036-15041.

Marsiglio, W. (1987). Adolescent fathers in the United States. *Family Planning Perspectives,* 19, 240-251.

Marsiglio, W. (1987). Adolescent males' orientation toward paternity and contraception. *Family Planning Perspectives,* 25, 22-31.

Martens, P. L. (1997). Immigrants, crime and criminal justice in Sweden. In M. Tonry (Ed) *Ethnicity, crime and immigration.* Chicago: University of Chicago Press.

Mason, D. A. & Frick, P. J. (1994). The heritability of psychopathic behavior . *Journal of Psychopathology & Behavioral Assessment,* 16, 301–323.

Matasha, E., Ntembelea, T., Mayaud, P. et al., (1998). Sexual and reproductive health among primary and secondary school pupils in Mwanza, Tanzania. *AIDS Care,* 10, 571-582.

Maticka-Tyndale, E., Godin, G., LeMay, G. et al. (1996). Canadian ethnocultural communities facing AIDS. *Canadian Journal of Public Health,* 87, S-38-43.

Matzopoulos, R. G., Thompson, M. Lou, & Myers, J. E. (2014). Firearm and nonfirearm homicide in 5 South African cities: a retrospective population-based study. *American Journal of Public*

Health, 104, 455-460.

Mau, W.C. & Lynn, R. (1999). Racial and ethnic differences in motivation for educational achievement in the United States. *Personality and Individual Differences*, 27,1091-1097.

Mazur, A. & Booth, A. (2014). Testosterone is related to deviance in male army veterans. *Biological Psychology*, 96, 72-76.

Mbagaya1, C., Oburu, P. & Bakermans-Kranenburg, M. J. (2013). Child physical abuse and neglect in Kenya, Zambia and the Netherlands: A cross-cultural comparison of prevalence, psychopathological sequelae and mediation by PTSS. *International Journal of Psychology*, 48, 95–107.

McCoy, W. & Edens, J. (2006). Do black and white youths differ in levels of psychopathy traits? A meta-analysis of the Psychopathy Checklist measures. *Journal of Consulting and Clinical Psychology*, 74, 386-392.

McCrae, R. R. (2002). NEO-PI-R data from 36 cultures. In R. R. McCrae & J. Allik (Eds) *The Five-Factor Model across Cultures*. New York: Kuwer.

McDermott, P. A. & Spencer, M. B. (1997). Racial and social class prevalence of psychopathology among school-age youth in the United States. *Youth and Society*, 28, 387–414.

McFarlane, C. P. (1999). *Reproductive Health Survey Jamaica 1997*. Atlanta, GA: US Dept. Health and Human Services..

McLean, P. E. (1995). Sexual behaviors and attitudes of high school students in the kingdom of Swaziland. *Journal of Adolescent Research*, 10, 400-420.

McNamara, P., Guadagnoli, E., Evanisko, M. J. et al. (1999). Correlates of support for organ donation among three ethnic groups. *Clinical Transplantation*, 13, 45–50.

McNulty, T. L., Bellair, P. E. & Watts, S. J. (2014). Neighborhood disadvantage and verbal ability as explanations of the Black–White difference in adolescent violence: Toward an integrated model. *Crime and Delinquency*, 59, 140-160.

Mealy, L. (1995). The sociobiology of sociopathy: an integrated evolutionary model. *Behavioral and Brain Sciences*, 18, 523-599.

Meekers, D. & Ahmed, G. (2000). Contemporary patterns of adolescent sexuality in urban Botswana. *Journal of Biosocial Science*, 32, 467-485.

Meisenberg, G. & Woodley, M. A. (2013). Global behavioral variation: A test of differential-K. *Personality and Individual Differences,* 55, 273-278.

Mensch, B. S. & Kandel, D. B. (1988). Underreporting of substance abuse in a national longitudinal youth cohort. *Public Opinion Quarterly*, 52, 100-124.

Messersmith, L. J., Kane, T. T. Odebiyi, A. I. et al. (2000). Who's at risk? Men's STD experience and condom use in southwest Nigeria. *Studies in Family Planning*, 31, 203-216.

Meston, C. M., Trapnell, P. D. & Gorzalka, B. B. (1996). Ethnic and gender differences in sexuality. *Archives of Sexual Behavior*, 25, 33-72

Miller, L. S., Klein, R. G. & Piacentini, J. (1995). The New York teacher rating scale for disruptive and psychopathic behavior. *Journal of the American Academy of Child and Adolescent Psychiatry*, 34, 359–370.

Minkov, M. (2014). The *K* factor, societal hypometropia, and national values: a study of 71 nations. *Personality and Individual Differences*, 66, 153-159.

Mischel, W. (1961a). Delay of gratification, need for achievement and acquiescence in another culture. *Journal of Abnormal and Social Psychology*, 62, 543–552.

Mischel, W. (1961b). Father absence and delay of gratification: cross-cultural comparisons. *Journal of Abnormal and Social Psychology*, 63, 116–124.

Mkandawire, P., Luginaah, I. & Dixon, J. (2013). Circumcision status and time to first sex among never-married young men in Malawi. *Aids and Behavior*, 17, 2123-2135.

Mnyika, K. S., Kvale, G. & Klepp, K. I. (1995). Perceived function of and barriers to condom use in Arusha and Kilimanjaro regions of Tanzania. *AIDS Care*, 7, 295-305.

Model, S., Fisher, G., & Silberman, R. (1999). Black Caribbeans in comparative perspective. *Journal of Ethnic and Migration Studies*, 25, 187–212.

Modood, T. & Berthoud, R. (1997). *Ethnic minorities in Britain*. London: Policy Studies Institute.

Moffit, T. E. (1993). Adolescence-limited and life-course-persistent psychopathic behavior: a developmental taxonomy. *Psychological Review*, 100, 674-701.

Moore, D. S. & Erickson, P. I. (1985). Age, gender and ethnic differences in sexual and contraceptive knowledge, attitudes and behaviors. *Family Community Health*, 8, 38-51.

Moore, K. A., Manlove, J., Glei, D. A. & Morrison, D. R. (1998). Non-marital school-age motherhood: family, individual and school characteristics. *Journal of Adolescent Research*, 13, 433–457.

Moore, K. A. & Stief, T. M. (1991). Changes in marriage and fertility behavior . *Youth and Society*, 22, 286–362.

Moore, M. J., Barr, E. M & Johnson, T. M. (2013). Sexual behaviors of middle school students: 2009 Youth Risk Behavior Survey results from 16 locations. *Journal of School Health*, 83, 61-80.

Moore, V. L. & Schwebel, A. I. (1993). Factors contributing to divorce: a study of race differences. *Journal of Marriage and Divorce*, 20, 123–135.

Moran, P. (1999). The epidemiology of psychopathic personality disorder. *Social Psychiatry and Psychiatric Epidemiology*, 34, 231–242.

Morey, L. C., (1991). *Personality Assessment Inventory*. Odessa, FL: PAR Inc.

Morgan, S. E. & Miller, J. K. (2002). Communicating about gifts of life: the effected of knowledge, attitudes, and altruism on behavior and behavioral intentions regarding organ donation. *Journal of Applied Community Research*, 30, 163-78.

Morris, A. & Reilly, J. (2003). *New Zealand National Survey of Crime Victims*. Wellington: Ministry of Justice.

Morris, L. (1988). Young adults in Latin America and the Caribbean: their sexual experience and contraceptive use. *International Family Planning Perspectives*, 14, 153-158.

Mosher, W. D. & Jones, J. (2010). Use of contraception in the United States. 1982-2008. *Vital and Health Statistics*, series 2, N0 29.

Mosli, H.A. (2003). Prostate cancer in Saudi Arabia in 2002. *Saudi Medical Journal*, 24, 573-581.

Muckle, G., Laflamme, D., Gagnon, J. et al (2011). Alcohol, smoking and drug use among Inuit women of childbearing age during pregnancy and the risk to children. *Alcohol Clinical Experimental Research*, 35, 1081–1091.

Munguti, K., Grosskurth, H., Newell, J. et al. (1997). Patterns of sexual behavior in a rural population in north-western Tanzania. *Social Science and Medicine*, 44, 1553-1561.

Munro, B. E. & Adams, G. R. (1978). Correlates of romantic love revisited. *Journal of Psychology*, 98, 211–214.

Munsch, J. & Wampler, P. S. (1993). Ethnic differences in early adolescents' coping with school stress. *American Journal of Orthopsychiatry*, 63, 633–646.

Murphy, J. M. (1976). Psychiatric labeling in cross-cultural perspective. *Science*, 191, 1019-1028.

Myers, L. & Levy, G. (1978). Description and prediction of the intractable inmates. *Crime and Delinquency*, 15, 214-228.

Nagy, C., Leal-Puente, L., Trainor, D. J. & Carlson, R. (1997). Mexican American adolescents' performance on the MMPI-A. *Journal of Personality Assessment*, 69, 205-214.

National Center for Health Statistics (1991). *Health United States*. Washington, D.C.: National Center for Health Statistics.

Navarro, AM. (1999). Smoking status by proxy and self-report: Rate of agreement in different ethnic groups. *Tobacco Control*, 8, 182-185.

Ndeki, S. S., Lepp, K. I. & Liga, G. R. (1994). Knowledge, perceived risk of AIDS and sexual behavior among primary school children in two areas of Tanzania. *Health Education Research,* 9, 133-138.

Nedopil, N., Hollweg, M., Hartman, J. & Jaser, R. (1998). Comorbidity of psychopathy with major mental disorders. In J.D. Cook, A. E. Forth & R. D. Hare (Eds) *Psychopathy: Theory, Research and Implications for Society.* Dortrecht, Netherlands: Kluwer Academic.

Neto, F., Mullet, E., Deschamps, J. et al. (2000). Cross-cultural variation in attitudes toward love. *Journal of Cross-cultural Psychology,* *31,* 626–635.

Neumann, C. S. & Hare, R. D. (2008). Psychopathic traits in a large community sample: links to violence, alcohol use, and intelligence. *Journal of Consulting & Clinical Psychology,* 76, 893-899.

Neumann, C. S. & Pardini, D. (2012). Factor structure and construct validity of the Self-Report Psychopathy (SRP) Scale and the Youth Psychopathic Traits Inventory (YPI) in Young Men. *Journal of Personality Disorders,* 26, 1-15.

Neumann, C. S., Schmitt, D. S., Carter, R., Embley, I. & Hare, R. D. (2012). Psychopathic traits in females and males across the globe. *Behavioral Sciences and the Law,* 30, 557-574.

Newby, J. H., McCarroll, J. E., Thayer, L. E., Norwood, A. E., Fullerton, C. S. & Ursano, R. J. (2000). Spouse abuse by black and white offenders in the US army. *Journal of Family Violence,* 15, 199–208.

Newman, J. P., Kosson, D. S. & Patterson, C. M. (1999). Delay of gratification in psychopathic and non- psychopathic offenders. *Journal of Abnormal Psychology,* 101,630-636.

New Zealand (1970, 1996). *Criminal Statistics.* New Zealand Statistical Office.

New Zealand (2007). *Drug use in New Zealand: analysis of the 2003 New Zealand health behaviors survey, drug use.* Wellington: Ministry of Health.

New Zealand Dept Corrections (2007). *Over-representation of Maori in the Criminal Justice System*. New Zealand Dept Corrections.

Niemcryk, S. J., Brawley, M., Kaufman, R. C. & Young, S. I. (1997). Motor vehicle crashes, restraint use and severity of injury in children in Nevada. *American Journal of Preventative Medicine*, 13, 109–114.

Nigg, J. T. & Goldsmith, H. H. (1994). Genetics of personality disorders. *Psychological Bulletin*, 115, 346–380.

Norman, L. R. (2001). Sexually transmitted disease symptoms in Jamaica. *West Indian Medical Journal*, 50, 203-208.

Nyborg, H. (2013). Migratory selection for inversely related covariant T-, and IQ-nexus traits: testing the IQ/T geo-climatic origin theory og the general trait covariance model. *Personality and Individual Differences*, 55, 267-272.

Nzewi, E. N. (1998). Use of the MMPI 2 with Nigerian students. *News and Profiles*, 9.

Oetting, E. R., Goldstein, G. S. & Beauvais, F. (1980). *Drug abuse among Indian children*. Fort Collins, CO: Western Behavioral Studies.

Oetting, E. R. & Beauvais, F. (1990). Adolescent drug use: Findings of national and local surveys. *Journal of Consulting and Clinical Psychology*, 58, 385-394.

Office of the Attorney General, State of Hawaii. (1987). *Crime in Hawaii*. Honoulu, HI.

Ontario (1996). *Report of the commission on systemic racism in the Ontario criminal justice system*. Toronto, Queen's Printer: Ministry of the Solicitor-General and Correctional Services.

Overpeck, M. D., Brenner, R. A., Trumble, A. C., Trifeletti, L. B. & Berendes, H. W. (1998). Risk factors for infant homicide in the United States. *New England Journal of Medicine*, 339, 1211-1216.

Pace, T.M., Robbins, R.R., Choney, S.K. et al. (2006). A cultural-contextual perspective on the validity of the MMPI-2 with American Indians. *Cultural Diversity and Ethnic Minority Psychology*, 12, 320-333.

Pagovich,O.(2004). Israel. In K.Malley-Morrison (Ed) *International Perspectives on Family Violence*. Mahwah, NJ: Lawrence Erlbaum.

Park, J. Y. & Johnson, R. C. (1984). Moral development in rural and urban Korea. *Journal of Cross Cultural Psychology*, 15, 35-46.

Paris, J. (2013). Antisocial and borderline personality disorder revisited. *Comprehensive Psychiatry*, 54, 321-325.

Parra, E. J, Marcini, A. & Akey, J. (1998). Estimating African-American admixture portions by use of population specific alleles. *American Journal of Human Genetics*, 63, 1839-1851.

Passamonti, L., Fairchild, G., Goodyer, I.M. et al. (2010). Neural abnormalities in early-onset and adolescence-onset conduct disorder. *Archives of General Psychiatry*, 67, 729-38.

Peacock, R.B., Day, P.A. & Peacock, T.D. (1999). Adolescent gambling on a great lakes Indian reservation. *Journal of Human Behavior in the Social Environment*, 2, 5-17.

Perez, P.R. (2012). The etiology of psychopathy: a neuropsychological perspective. *Aggression and Violent Behavior*, 17, 519-522.

Perkins, J. J., Sanson-Fisher, W. R. & Blunden, S. (1994). The prevalence of drug use in Aboriginal communities. *Addiction*, 89, 1319–1331.

Petersilia, J. & Honig, P. K. (1980). *The prison experience of career criminals.* Santa Monica, CA: Rand.

Peterson, J. L., Coates, T. J., Catania, L. et al., (1992). High risk sexual behavior and condom use among gay and bisexual African-American men. *American Journal of Public Health*, 82, 1490-1494.

Peterson, L. S., Oakley, D., Potter, L. S. & Darroch, J. E. (1998). Women's efforts to prevent pregnancy: consistency of oral contraceptive use. *Family Planning Perspectives*, 30, 19–23.

Peterson, S. M. (1997). Are young black men really less willing to work? *American Sociological Review*, 62, 605-613.

Pham, H. & Spigner, C. (2004). Knowledge and opinions about organ donation and transplantation among Vietnamese Americans in Seattle, Washington: a pilot study. *Clinical Transplantation*, 18, 707–715.

Philbrick, J. L., Thomas, F. F., Cretser, G. A. & Leon, J. J. (1988). Sex differences in love attitudes of black students. *Psychological Reports*, 62, 414.

Philips (1996). *World Atlas*. London: Chancellor Press.

Piffer, D. (2011). Race differences in maturation: Some data for the age of voice breaking in Europeans, South Asians and sub-Saharan Africans. *International Journal of Anthropology*, 26,253-255.

Pinel, P. (1801). *Traité médico-philosophique sur l'aliénation mentale*. Paris.

Pink, B. & Allbon, P. (2008). *The health and welfare of Australia's Aboriginal and Torres Strait Islander peoples 2008*. Canberra: Australian Bureau of Statistics.

Piquero, A. R., Jennings, W. G., Diamond, B., & Reingle, J. M. (2015). A systematic review of age, sex, ethnicity, and race as predictors of violent recidivism. *International Journal of Offender Therapy and Comparative Criminology*, 59, 5-26.

Polednak, A.P. (1989). *Racial and Ethnic Differences in Disease*. Oxford: Oxford University Press.

Pomare, E., Keefe-Ormsby, V., Ormsby, C. & Pearce, N. (1995). *Hauora. Maori Standards of Ill-health III*. Wellington, NZ: Te Ropu Rangahau Hauora a Eru Pomare.

Porteus, S. D. & Babcock, M. E. (1926).*Temperament and Race*. Boston: Badger.

Porter, B. E. & England, K. J. (2000). Predicting red-light running behavior : a traffic safety study in three urban settings. *Journal of Safety Research*, 31, 1–8.

Preston, D. (1979). *A moral education program conducted in the health and physical education curriculum*. PhD thesis, University of Georgia.

Price, R. K., Wada, K. & Murray, K. S. (1995). Protective factors for drug abuse. In R. K. Price, B. M. Shea & H. N. Mookherjee (Eds)

Social Psychiatry Across Cultures. New York: Plenum.

Price-Williams, D. R. & Ramirez, M. (1974). Ethnic differences in delay of gratification. *Journal of Social Psychology*, 93, 23–30.

Pritchard, J. C. (1835). *A treatise on insanity and other diseases affecting the mind.* Philadelphia: Harwell, Barrington and Harwell.

Productivity Commission (1999). *Australia's Gambling Industries.* Final Report No. 10. AusInfo, Canberra.

Prostate Cancer UK (2013). *Working out the risk of prostate cancer in Black men.* London.

Purdie, N. & McCrindle, A. (2004). Measurement of self-concept among indigenous and non-indigenous Australian students. *Australian Journal of Psychology*, 56, 50–62.

Putnam-Hornstein, E., Needell, B., King, B. & Johnson-Motoyamac, M. (2013). Racial and ethnic disparities: A population-based examination of risk factors for involvement with child protection services. *Child Abuse and Neglect*, 37, 33-46.

Puzone, C.A., Saltzman, L.E., Kesno, M-J., Thomson, M.P. & Mercy, J.A. (2000). National trends in intimate partner homicide: United States, 1976-1995: erratum. *Violence against Women*, 6, 1179-1184.

Quadagno, Sly, D. F., Harrison, D. F. et al. (1998). Ethnic differences in sexual decisions and sexual behavior. *Archives of Sexual Behavior*, 27, 57-75.

Quinlan, R. J. (2008). Human pair-bonds: evolutionary functions, ecological variation, and adaptive development. *Evolutionary Anthropology*, 17, 227-238.

Raine, A. (1993). *The psychopathology of crime.* San Diego, CA: Academic Press.

Raine, A. (2013). *The Anatomy of Violence.* New York: Random House.

Raine, A., Venables, P. H. & Mednick, S. A. (1997). Low resting heart rate at age 3 years predisposes to aggression at age 11 years: evidence from the Mauritius Child Health Project. *Journal of the American*

Academy of Child and Adolescent Psychiatry, 36, 1457–1464.

Raine, T. R., Jenkins, R., Aarons, S.J. et al. (1999). Socio-demographic correlates of virginity in seventh-grade black and Latino students. *Journal of Adolescent Health*, 24, 304-312.

Raji, T. A. & Adegboye, A. A. (1993). Condom acceptance and husbands of teenage-wives in northern Nigeria. *VIIIth Congress on AIDS in Africa, Marrakesh.*

Raley, R. J. (1996). A shortage of marriageable men? A note on the role of cohabitation in black-white differences in marriage rates. *American Sociological Review*, 61, 973–983.

Ram, B. & Ebanks, G. E. (1973). Stability of unions in Barbados. *Social Biology*, 20, 143–153.

Ramirez, J. (1983). Race and the apprehension of inmate misconduct. *Journal of Criminal Justice,* 11, 413-427.

Rede Justiça Criminal (2016). Human Rights and Criminal Justice in Brazil. 3rd Cycle of the Universal Periodic Review. October 6, 2016. http://www.global.org.br/wp-content/uploads/2016/10/RPU-Sistema-Prisional.pdf. Accessed February 15, 2018.

Rest, J. R. (1979). *Development in judging moral issues.* Minneapolis, MN: University of Minnesota Press.

Rest, J. R. (1986). *Moral development.* Westport, CT: Praeger.

Reti, I. M., Xu, J. Z., Yanofski, J., McKibben, J. et al. (2011). Monoamine oxidase A regulates antisocial personality in whites with no history of physical abuse. *Comprehensive Psychiatry*, 52,188–194.

Richter, D. L., Valois, R. F., McKeown, R. E. & Vincent, M. L. (1993). Correlates of condom use and number of sexual partners among high school students. *Journal of School Health*, 63, 91-96.

Roberts, J. V. & Melchers, R. (2003). The incarceration of aboriginal offenders: Trends from 1978 to 2001. *Canadian Journal of Criminology and Criminal Justice*, 45, 211-242.

Robin, R. W., Greene, R. L. & Albaugh, B. (2003). Use of the MMPI–2 in American Indians: I. Comparability of the MMPI–2 between two tribes and with the MMPI–2 normative group. *Psychological Assessment*, 15, 351-359.

Robins, L. N., Helzer, J. E., Weissman, M. M. & Orvaschel, H. (1984). Lifetime prevalence of specific psychiatric disorders in three sites. *Archives of General Psychiatry*, 41, 949-958.

Robins, L.N. & Regier, D.A. (1991). *Psychiatric disorders in America: the epidemiological catchment area study.* New York: Free Press.

Robins, L. Tipp, J. & Przybeck, T. (1991). Psychopathic personality. In L. Robins & D. Regnier (Eds) *Psychiatric disorders in America.* New York: Free Press.

Rocca, C. H. (2012). Do racial and ethnic differences in contraceptive knowledge explain disparities in method use? *Perspectives on Sexual and Reproductive Health*, 14, 150-158.

Rodham, K., Hawton, K., Evans, E. & Weatherall, R. (2005). Ethnic and gender differences in drinking, smoking and drug taking among adolescents in England: a self-report school-based survey of 15 and 16 year olds. *Journal of Adolescence,* 28, 63-73.

Rolfs, R.T. & Nakashima, A. K. (1990). Epidemiology of primary and secondary syphilis in the United States1981 through 1989. *Journal of the American Medical Association*, 264, 1432-1437.

Rosenthal, D., Moore, S. & Brumen, I. (1990). Ethnic group differences in adolescents' responses to AIDS. *Australian Journal of Social Issues*, 25, 330-239.

Ross, R., Bernstein, L., Judd, H., Pike, M. & Henderson, B. (1986). Serum male testosterone levels in healthy young black and white men. *Journal of the National Cancer Institute*, 76, 45-48.

Rowe, D.C. (2002). IQ, birth weight, and number of sexual partners in White, African American, and Mixed Race adolescents. *Population and Environment*, 23, 513-524.

Rubens, A. J. (1996) Racial and ethnic differences in students' attitudes and behavior toward organ donation. *Journal of the National Medical Association*, 88, 417–21.

Rushton, J. P. (2000). *Race, evolution and behavior: A life history perspective*. New Brunswick, NJ: Transaction.

Rushton, J.P. & Bogaert, A.F. (1987). Race differences in sexual behavior: testing an evolutionary hypothesis. *Journal of Research in Personality*, 21, 529-551.

Rushton, J.P. & Templer, D. (2009). National IQ, national income, skin color, and crime. *Intelligence*, 37, 341-346

Rutter, M., Yule, W., Berger, M., Yule, B., Morton, J. & Bagley, C. (1974). Children of West Indian Immigrants -1. Rates of behavioral deviance and of psychiatric disorders. *Journal of Child Psychology & Psychiatry*, 15, 241–262.

Sabogal, F., Perez-Stable, E. J., Otero-Sabogak, R. & Hiatt, R. A. (1995). Gender, ethnic, and acculturation differences in sexual behaviors: Hispanic and non-Hispanic white adults. *Hispanic Journal of Behavioral Sciences*, 17, 139-159.

Saha, N. & Samuel, A. (1987). A genetic study of blacks from Trinidad. *Human Heredity*, 37, 365-370.

Salekin, R. T., Rogers, R. & Sewell, K. W. (1996). A review and meta-analysis of the Psychopathy Checklist and Psychopathy Checklist–Revised: Predictive validity of dangerousness. *Clinical Psychology—Science and Practice*, 3, 203-215.

Sampson, .R. & Bartusch, D. (1998). Legal cynicism and (subcultural?) tolerance of deviance: the neighborhood context of racial differences. *Law and Society Review*, 32, 777-804.

Samuels, H. P. (1997). The relationships among selected demographic variables and conventional and unconventional sexual behaviors among black and white heterosexual men. *Journal of Sex Research*, 34, 85-92.

Samuels, J., Eaton, W. W. & Bienvenu, J. (2002). Prevalence and correlates of personality disorders in a community sample. *British Journal of Psychiatry*, 180, 536-542.

Sanner, M.A. (1998). Giving and taking—to whom and from whom? People's attitudes toward transplantation of organs and tissue from different sources. *Clinical Transplantation*, 12, 530–37.

Santinelli, J. S., Brener, N. D., Lowry, B. & Robin, L. S. (1998). Multiple sexual partners among US adolescents and young adults. *Family Planning Perspectives*, 30, 271–275.

Santinelli, J. S., Lowry, B., Brener, N. D. & Robin, L. (2000). The association of sexual behaviors with socioeconomic status, family structure and race/ethnicity among US adolescents. *American Journal of Public Health*, 90, 1582-1587. .

Sapienza, P., Maestripieri, D. & Zingales, L. (2009). Gender differences in financial risk aversion and career choices are affected by testosterone. *Proceedings of the National Academy of Sciences*, 106: 15268-15273.

Schiff, M., & Becker, T. (1996). Trends in motor vehicle traffic fatalities among Hispanics, non-Hispanic whites and American Indians in New Mexico, 1958–1990. *Ethnicity and Health*, 1 283-291.

Schmitt, D. P., Youn, G., Bond, B. et al. (2009). When will I feel love? The effects of culture, personality, and gender on the psychological tendency to love. *Journal of Research in Personality*, 43, 830–846.

Schmitt, D. F. (2003). Universal sex differences in desire for sexual variety. *Journal of Personality & Social Psychology*, 85, 85-104.

Schuster, M. A., Bell, R. M., Nakajima, G. A. & Kanouse, D. E. (1998). The sexual practices of Asian and Pacific Islander high school students. *Journal of Adolescent Health*, 23, 221-231.

Scott, E. L., Eng, W. & Heimberg, R. G. (2002). Ethnic differences in worry in a nonclinical population. *Depression and Anxiety*, 15, 79-82.

Schneider, H. J. (1992). Life in societal no-man's land: Aboriginal crime in cenral Australia. *International Journal of Offender Therapy and Comparative Criminology*, 36, 5-19.

Schulsinger, F. (1972). Psychopathy: heredity and environment. *International Journal of Mental Health*, 1, 190-206.

Schuster, M. A., Bell, R. M., Nakajjima, G. A. & Kanouse, D. E. (1998). The sexual practices of Asian and Pacific Islander high school students. *Journal of Adolescent Health,* 23, 221-231.

Schwarz, D. F., Grisso, J. A., Miles, C. G., Holmes, J. H., Wishner, A. R. & Sutton, R. L. (1994). A longitudinal study of injury morbidity in an African American population. *Journal of the American Medical Association,* 271, 755-760.

Scott, E. L., Eng, W. & Heimberg, R. G. (2002). Ethnic differences in worry in a nonclinical population. *Depression and Anxiety, 15,* 79–82.

Seagull, A. A. (1966). Sub-patterns of gratification and choice within samples of Negro and white children. *Papers of the Michigan Academy of Science, Arts and Letters,* 51, 345–351.

Segal, B. (1998). Drinking and drinking related problems among Alaska natives. *Alcohol Health and Research World,* 22, 276–280.

Segal, N.L. (2012). *Born together-reared apart.* Cambridge, MA: Harvard University Press.

Seidman, S. N. & Aral, S. (1992). Subpopulation differences in STD transmission. *American Journal of Public Health,* 82, 1297.

Selten, J.P. & Sijben, N. (1994) First admission rates for schizophrenia in immigrants to the Netherlands. *Social Psychiatry and Psychiatric Epidemiology,* 29, 71-77.

Sher, L. (2014). Testosterone and homicidal behavior. *Australian and New Zealand Journal of Psychiatry,* 48, 290.

Shi, L. (1999). Experience of primary care by racial and ethnic groups in the United States. *Medical Care,* 37, 1068-1077.

Shin, D., Hong, L. & Waldron, I. (1999). Possible causes of socioeconomic and ethnic differences in seat belt use among high school students. *Accident Analysis and Prevention,* 31, 485-496.

Sikkema, K. J., Brondino, M. J., Anderson, E. S. & Gore-Felton, C. (2004). HIV risk behavior among ethnically diverse adolescents living in low-income housing developments. *Journal of Adolescent Health,* 35, 141-150.

Skeem, J. L., Edens, J. F., Sanford, G. M & Colwell, L. H. (2003).

Psychopathic personality and racial/ethnic differences: a reply to Lynn (2002). *Personality and Individual Differences*, 35, 1439-1462.

Skeem, J. L., Edens, J. F., Camp, J. & Colwell, L. H. (2004). Are there ethnic differences in levels of psychopathy? A meta-analysis. *Law and Human Behavior*, 28, 505-527.

Skeem, J. L., Mulvey, E. & Grisso,T. (2003). Applicability of traditional and revised models of psychopathy to the Psychopathy Checklist: Screening Version. *Psychological Assessment*, 15, 41-55.

Skeem, J. L., Poythress, N., Edens, J. F., Lilienfeld, S. O. & Cale, E. M. (2003). Psychopathic personality or personalities? *Aggression & Violent Behavior*, 8, 513-546.

Sijtsema, J. J., Baan, L., & Bogaerts, S. (2014). Associations between dysfunctional personality traits and intimate partner violence in perpetrators and victims. *Journal of Interpersonal Violence*, 29, 2418-2438.

Silberg, J., Meyer, J., Pickles, A., Simonoff, E., Eaves, L., Hewitt, J., Maes, H. & Rutter, M. (1996). Heterogeneity among juvenile psychopathic behaviors: findings from the Virginia twin study of adolescent behavioral development. In G. Rock & J. Goode (Eds) *Genetics of criminal and psychopathic behaviors*. Chichester, UK: Wiley.

Sinclair, H. & Feigenbaum, J. (2012). Trait emotional intelligence and borderline personality disorder. *Personality and Individual Differences*, 52, 674-679.

Slap, G.B., Lot, L., Huang, B. et al. (2003). Sexual behavior of adolescents in Nigeria: cross sectional survey of secondary school students. *British Medical Journal*, 326, 15-21.

Smith, D. F. (1997). Ethnic origins, crime and criminal justice in England and Wales. In M. Tonry (Ed) *Ethnicity, crime and immigration*. Chicago: University of Chicago Press.

Smith, M.G. (1984). *Culture, Race, and Class in the Commonwealth Caribbean*. Jamaica: University of the West Indies Press.

Smith, P.H., Thornton, G.E., DeVellis, R. et al. (2002). A population-based study of the prevalence and distinctiveness of battering, physical assault and sexual assault in intimate relationships. *Violence against Women*, 8, 1208-1232.

Snipp, S. M. (1991). *American Indians: first of this land*. New York: Russell Sage.

Sonenstein, F.L., Ku, L., Lindberg, L.D. et al. (1998). Changes in sexual behavior and condom use among teenage males, 1988 to 1995. *American Journal of Public Health*, 88, 956-959.

Sonenstein, F. L., Pleck, J. H. & Ku, L. C. (1989). Sexual activity, condom use and Aids awareness among adolescent males. *Family Planning Perspectives*, 21, 152–158.

Sonenstein, F. L., Pleck, J. H. & Ku, L. C. (1991). Levels of sexual activity among adolescent males in the United States. *Family Planning Perspectives*, 23, 162–167.

Sorensen, J. & Wrinkle, R. D. (1996). No hope for parole: Disciplinary infractions among death-sentenced and life without parole inmates. *Criminal Justice & Behavior* , 23, 542-552l

Sourifu (1999). *FY1999 Annual report on the State of the Formation of a Gender Equal Society*. Tokyo: Prime Mimister's Office.

South, S.C. (2013). Behavior genetics of personality disorders. *Personality Disorders –Theory, Research & Treatment*, 4, 270-283.

South, S. J. (1993). Racial and ethnic differences in the desire to marry. *Journal of Marriage and the Family*, 55, 357-370.

South, S. J. & Lloyd, K. M. (1992). Marriage opportunities and family formation: further implications for imbalanced sex ratios. *Journal of Marriage and the Family*, 54, 440-451.

Sowell, T. (1978). *American ethnic groups*. Washington, DC: Urban Institute.

Speizer, I.S. & Yates, A.J. (1998). Polygyny and African couple research. *Population Research and Policy Review*, 17, 551-570.

Spigner, C., Weaver, M., Cárdenas, V., & Allen, M. D. (2002). Organ donation and transplantation: Ethnic differences in knowledge and opinions among urban high school students. *Ethnicity and Health,* 7,

87-101.

Spring, C., Blunden, D. S., Greenberg, L. M. & Yellin, A. M. (1977). Validity and norms of a hyperactivity rating scale. *Journal of Special Education*, 11, 313–321.

Stanik, C.E., McHale, S.M. & Crouter, A.C. (2013). Gender dynamics predict changes in marital love among African-American couples. *Journal of Marriage and the Family*, 75, 795-807.

St. Louis, M.E., Conway, G.A., Hayman, C.R. et al. (1991). Human immunodeficiency virus infection in disadvantaged adolescents. *Journal of the American Medical Association*, 266, 2387-2391.

Staples, R. & Johnson, L. B. (1993). *Black families at the cross roads: challenges and prospects*. San Francisco: Jossey-Bass.

Statistic Canada (2009). *Incident-based Uniform Crime Reporting Survey, 2006 to 2008*. Ottawa: Canadian Centre for Justice Statistics.

Statistics Canada (2012a). *Births 2009*. Ottawa: Statistics Canada.

Statistics Canada (2012b). *Canadian Vital Statistics, Birth Database*. Ottawa: Statistics Canada.

Steering Committee for the Review of Government Service Provision (2009). *Overcoming Indigenous disadvantage: Key indicators 2009 report*. Canberra: Productivity Commission.

Sterelny, K. (2013). Life in interesting times: cooperation and collective action in the Holocene. In K. Sterelny, R. Joyce, B. Calcott & B. Fraser (Eds) *Cooperation and its evolution*. Cambridge, MA: MIT Press.

Stinchfield, R. (2000). Gambling and correlates of gambling among Minnesota public school students. *Journal of Gambling Studies*, 16, 153-173.

Stephen, E. H., Rindfuss, R. R. & Bean, F. D. (1988). Racial differences in contraceptive choice: complexity and implications. *Demography*, 25, 53-70.

Stevens, M., & Young, M. (2009a). Betting on the evidence: Reported gambling problems among the indigenous population of the Northern Territory. *Australian and New Zealand Journal of Public Health*, 33, 556–565.

Stevens, M., & Young, M. (2009b). Independent correlates of reported gambling problems amongst indigenous Australians. *Social Indicators Research*, 98, 147–166.

Stockman, J. K., Lucea, M. B., Bolyard, R. et al. (2014). Intimate partner violence among African American and African Caribbean women: prevalence, risk factors, and the influence of cultural attitudes. *Global Health Action*, 7, 24772.

Straus, M. A., Gelles, R. J. & Steinmetz, S. K. (1980). *Behind closed doors: violence in the American family*. New York: Doubleday.

Strauss, M.A. & Smith, C. (1990). Violence in Hispanic families in the United States. In M.A. Strauss & R. J. Gelles (Eds) *Physical Violence in American Families*. New Brunswick, NJ: Transaction.

Substance Abuse and Mental Health Services Administration. (2012). *Results from the 2011 National Survey on Drug Use and Health: Summary of National Findings*, NSDUH Series H-44, HHS. Publication No. (SMA) 12-4713. Rockville, MD: Substance Abuse and Mental Health Services Administration.

Sullivan, M. & Grossman, D. C. (1999). Hospitalization for motor vehicle injuries among American Indians and Alaska Natives in Washington. *American Journal of Preventative Medicine*, 17, 38-42.

Summerfield, A. (2011). *Children and young people in custody, 2010-2011*. HM Inspectorate of Prisons: Youth Justice Board.

Swenson, I., Erickson, D., Ehlinger, E. et al. (1989). Fertility, menstrual characteristics, and contraceptive practices among white, black and southeast Asian refugee adolescents. *Adolescence*, 24, 647-654.

Taylor, D. J., Chavez, G. F., Adams, E. J., Chabra, A. & Shah, R. S. (1999). Demographic characteristics in adult paternity for first births to adolescents under 15 years of age. *Journal of Adolescent Health*, 24, 251–258.

Taylor, J. (2005). *The Color of Crime*. Oakton, VA: New Century

Foundation.

Taylor, J. & Whitney, G. (1999). US racial profiling in the prevention of crime: is there an empirical basis? *Journal of Social, Political and Economic Studies*, 24, 485-509.

Templer, D.I. (2013). Rushton: the great theoretician and his contribution to personality. *Personality and Individual Differences*, 55, 243-246.

Tomasello, M., Melis, A.,Tennie, C., Wyman, E. & Herrmann, E. (2012). Two key steps in the evolution of human cooperation. *Current Anthropology*, 53, 673-692.

Thompson, L. A. (1989). *Romans and Blacks.* Norman: University of Oklahoma Press.

Thompson, L. M. (1952). Indian immigration into Natal, 1860–72. *Archives Yearbook*, 2, 1-10.

Thomson, J. D. S (2004). A murderous legacy: Coloured homicide trends in South Africa. *SA Crime Quarterly*, 7, 9-14.

Thornberry, T. P., Smith, C. A. & Howard, G. J. (1997). Risk factors for teenage fatherhood. *Journal of Marriage and the Family*, 59, 505-522.

Thornton, J.D., Wong, K.A. & Cardenas, V. (2006) Ethnic and racial differences in willingness among high school students to donate organs. *Journal of .Adolescent Health*, 39, 266-276.

Timbruck, R. E. & Graham, J. R. (1994). Ethnic differences on the MMPI-2. *Psychological Assessment*, 6, 212-217.

Tippett, N., Wolke, D. & Platt, L. (2013). Ethnicity and bullying involvement in a national UK youth sample. *Journal of Adolescence*, 36, 639–649.

Ting, I. (2011). Aboriginal crime and punishment. *Crikey*, 15 Dec.

Tizard, B., Blntchford, P., Burke, J., Farquhar, C. & Plewis, J. (1988). *Young children at school in the inner city.* Mahwah, NJ: Lawrence Erlbaum.

Tomasello, M., Melis, A., Tennie, C., Wyman, E. & Herrmann,

E. (2012). Two key steps in the evolution of human cooperation. *Current Anthropology*, 53, 673-692.

Tonry, M. (1994). Racial disproportion in US prisons. *British Journal of Criminology*, 34, 97-115.

Torgersen, S., Kringlen, E. & Cramer, V. (2001). The prevalence of personality disorders in a community sample. *Archives of General Psychiatry*, 58, 590-596.

Tournier, P. (1997). *Nationality, crime and criminal justice in France*. In M. Tonry (Ed) *Ethnicity, crime and immigration*. Chicago: University of Chicago Press.

Tracy, P. E., Wolfgang, M. H. & Figlio, R. M. (1990). *Delinquency in two birth cohorts*. New York: Plenum.

Transparency International (2013). *Corruption Perceptions Index, 2012*. https://www.transparency.org/cpi2012/results. Accessed February 15, 2018.

Tremblay, R. E., Schaal, B., Boulerice, B., Arseneault, L., Soussignan, R.G., Paquette, D. & Laurent, D. (1998). Testosterone, physical aggression, dominance and physical development in early adolescence. *International Journal of Behavioral Development*, 22, 753-777.

Trent, K. & South, S. J. (1992). Socio-demographic status, parental background, childhood family structure and attitudes towards family formation. *Journal of Marriage and the Family*, 54, 427–439.

Tucker, S. K. (1991). The sexual and contraceptive socialization of black adolescent males. *Public Health Nursing*, 8, 105-111.

Turner, R. J. & Lloyd, D. A. (2004). Stress burden and the lifetime incidence of psychiatric disorder in young adults. *Archives of General Psychiatry*, 61, 481-488.

Tuvblad, C., Bezdjian, S., Raine, A. & Baker, L. A.(2014). The heritability of psychopathic personality in 14- to 15-year-old twins: A multirater, multimeasure approach. *Psychological Assessment*, 26,704-716.

Twenge, J. M. & Crocke, J. (2002). Race and self-esteem: meta-analysis comparing whites, blacks, Hispanics, Asians and American

Indians and a comment on Gray-Little & Hafdahl (2000). *Psychological Bulletin*, 128, 171-408.

UNAIDS (2000). *Report on the Global HIV/AIDS epidemic*. Geneva: UNAIDS.

UNAIDS (2009). *AIDS Epidemic Update*. Geneva: UNAIDS.

Unz, R. (2010). His-panic. *American Conservative*, 1 March.

U.S. Dept Education (2011). *Revealing New Truths About Our Nation's Schools*. Washington, D.C.: US Dept Education.

U.S. Department of Health and Human Services (1996). *Child maltreatment*. Washington, DC: Government Printing Office.

Uthman, O. A., Lawoko, S. & Moradi, T. (2009). Factors associated with attitudes towards intimate partner violence against women: a comparative analysis of 17 sub-Saharan countries. *BMC International Health & Human Rights*, 9,14-29

Valois, R. F., MCKeown, R. E., Garrison, C. Z. & Vincent, .L. (1995). Correlates of aggressive and violent behaviors among public high school adolescents. *Journal of Adolescent Health*, 16, 26-34.

Valois, R. F., Oeltmann, J. E., Waller, J. & Hussey, J. R. (1999). Relationship between number of sexual intercourse partners and selected health risk behaviors among public high school adolescents. *Journal of Adolescent Health*, 25, 328–335.

Van Laar, C. (2000). The paradox of low academic achievement but high self esteem in African American students: an attributional account. Educational *Psychology Review*, 12, 33-61.

Van Oort, F. V. A., Joung, I. M. A. & Mackenbach, J. P. (2007). Development of ethnic disparities in internalising and externalising problems from adolescence into young adulthood. *Journal of Child Psychology and Psychiatry*, 48, 176-184.

Vars, F. E. & Bowen, W. G. (1998). Scholastic aptitude test scores, race, and academic performance in selective colleges and universities. In C. Jencks and M. Phillips (Eds) *The black-white test score* gap. Washington, DC: Brookings Institution.

Vaughn, M. G., Maynard, B. R., Salas-Wright, C. P. et al. (2013). Prevalence and correlates of truancy in the US. *Journal of Adolescence*, 36, 767-776.

Vernon, P. E. (1969). *Intelligence and Cultural Environment*. London: Methuen.

Vernon, P. E. (1982). *The Abilities and Achievements of Orientals in North America*. New York: Academic Press.

Viding, E., Fontaine, N. M. & Larson, H. (2013). Quantitative genetic studies of psychopathic traits in minors. In K. A. Kiehl & W. S. Sinncott-Armstrong (Eds) *Handbook on Psychopathy and Law*. Oxford: Oxford University Press.

Visser, B.A., Pozzebon, .A., Bogaert, A. F. & Ashton, M. C. (2010). Psychopathy, sexual behavior, and esteem: it's different for girls. *Personality and Individual Differences*, 48, 833-838.

Vitelli, R. (1996). Prevalence of childhood conduct and attention deficit hyperactivity disorders in adult maximum security inmates. *International Journal of Offender Therapy and Comparative Criminology*, 40, 263–271.

Vives-Cases, C., Gil-Gonzalez, D., Plazaola-Castano, J., Montero-Pinar, M. I., Ruiz-Perez, I. & Escriba-Aguir, V. (2009). Violencia de genero en mujeres inmigrantes y espanolas: Magnitud, respuestas ante el problema y politicas existentes [Gender-based violence against immigrant and Spanish women: Scale of the problem, responses and current policies]. *Gaceta Sanitaria, 23* (Suppl.1), 100-106.

Voas, R. B., Wells, J. K., Lestina, D. C. et al. (1998). Drinking and driving in the United States: the 1996 national roadside survey. *Accident Analysis and Prevention*, 30, 267-275.

Volberg, R. A. & Abbott, M .A. (1997). Ethnicity and gambling: gambling and problem gambling among indigenous peoples. *Substance Use and Misuse*, 32, 1525-1538.

Volkwein, J. F., Szelest, B. P., Cabrera, A. F. & Napierski-Prancl, M. R. (1998). Factors associated with student loan default among different racial and ethnic groups. *Journal of Higher Education*, 69, 206–237.

Vung, N.D, Ostergren, P.O. & Krantz, G. (2008). Intimate partner violence against women in rural Vietnam-different socio-demographic factors are associated with different forms of violence. *BMC Public Health*, 8, 55-65.

Walls, M., Hartshorn, K. J. S. & Whitbeck, L. B. (2013). North American indigenous adolescent substance abuse. *Addictive Behaviors*, 38, 2103-2110.

Wang, L. & Graddy, E. (2008). Social capital, volunteering, and charitable giving. *Voluntas*, 19, 23–42.

Warner, J. T. & Pleeter, S. (2001). The personal discount rate: Evidence from military downsizing programs. *The American Economic Review*, 91, 33-53.

Warren, C. W., Santelli, J. S., Everett, S. A. & Kann, L. (1998). Sexual behavior among U.S. high school students, 1990–1995. *Family Planning Perspectives*, 30, 170–176.

Weissman, M. M. (1993). The epidemiology of personality disorders: An update. *Journal of Personality Disorders*, Supplement, 44-62.

Wellings, K., Field, J., Johnson, A. M. & Wadsworth, M. (1994). *Sexual behavior in Britain*. London: Penguin.

Wells, E. J., Bushnell, J. A. & Hornblow, A. R. (1989). Christchuch psychiatric epidemiology study. *Australian and New Zealand Journal of Psychiatry*, 23, 315-326.

Wells, J. K., Williams, A. F. & Farmer, C. M. (2002). Seat belt use among African Americans, Hispanics and Whites. *Accident Analysis and Prevention*, 34, 523-529.

Welte, J. W., Barnes, G., Wieczorek, W. et al. (2001). Alcohol and gambling pathology among U.S. adults: prevalence, demographic patterns and comorbidity. *Journal of Studies of Alcoholism*, 62,706–712.

Westermann, D. (1939). *The African Today and Tomorrow*. London: Methuen.

Westermayer, J. (1977). Cross-racial foster home placement among Native American psychiatric patients. *Journal of the National Medical*

Association, 69, 231–236.

White, B.A. (2014). Who cares when nobody is looking? Psychopathic traits and empathy in prosocial behaviors. *Personality and Individual Differences,* 56, 116-121.

WHO. (2002). *World Report on Violence and Health.* Geneva: World Health Organization.

Whitman, D.S., Kraus, E. & van Rooy, D.L. (2014). Emotional intelligence among black and white job applicants. *International Journal of Selection and Assessment,* 22, 199-210.

Widiger, T.A. & Lynam, D.R. (1998). Psychopathy and the five factor model of personality. In T. Millon & E. Simonsen (Eds) *Psychopathy: Psychopathic, criminal and violent behavior.* New York: Guildford Press.

Wiederman, M. W. (1997). Extramarital sex: prevalence and correlates in a national survey. *Journal of Sex Research,* 34, 167–174.

Williams, C. D. (1938). Child health in the Gold Coast *Lancet,* 1,97-106.

Williams, D.R ., Herman, A., Stein, D. J. et al. (2008). Twelve month mental disorders in South Africa. *Psychological Medicine,* 38, 211-220.

Wilson, D. S. (2014). *Does Altruism Exist?* New Haven: Yale University Press.

Wilson, E. O. (1975). *Sociobiology: The new synthesis.* Cambridge, MA: Harvard University Press.

Wilson, W. J. (1987). *The truly disadvantaged.* Chicago: University of Chicago Press.

Wingo, R., Chadiha, L. A. Vargas, A. & Mosley, M. (1998). Prostate cancer and psychosocial concerns in African American men. *Health and Social Work,* 28, 302-311.

Wolfgang, M. E., Figlio, R. M. & Sellin, T. (1972). *Delinquency in a birth cohort.* Chicago: University of Chicago Press.

Wong, M. M., Klingle, R. S., & Price, R. K. (2004). Alcohol, tobacco, and other drug use among Asian American and Pacific Islander Adolescents in California and Hawaii. *Addictive Behaviors,* 29, 127-141.

Woodward, A. T., Taylor, R. J., Bullard, K. M., Aranda, M. P., Lincoln, K. D. & Chatters, L. M. (2012). Prevalence of lifetime DSM-IV affective disorders among older African Americans, Black Caribbeans, Latinos, Asians and Non-Hispanic White people. *International Journal of Geriatric Psychiatry*, 27, 816-827.

Wooldridge, J.D., Griffin, T. & Pratt, T. (2001). Considering hierarchical models for research on inmate behavior : Predicting misconduct with multilevel data. *Justice Quarterly*, 18, 203-231.

Wright, J.P., Morgan, .A., Coyne, M.A., Beaver, K.M. & Barnes, J.C. (2014). Prior problem behavior accounts for the racial gap in school suspensions. *Journal of Criminal Justice,* 42, 257-256.

Wu, Z., Noh, S., Kaspar, V. & Schmele, C. M. (2003). Race, ethnicity and depression in Canadian society. *Journal of Health & Social Behavior*, 44, 426-441.

Wyatt, G.E. (1991). Examining ethnicity versus race in AIDS related sex research. *Social Science and Medicine*, 33, 37-45.

Yao, K.-N., Solanto, M. V. & Wender, E. H. (1988). Prevalence of hyperactivity among newly immigrated Chinese-American children. *Developmental and Behavioral Pediatrics*, 9, 367–373.

Yeung, I., Kong, S.H. & Lee J. (2000). Attitudes towards organ donation in Hong Kong. *Social Science and Medicine,* 50, 1643-1654.

Young, J. W. (1991). Improving the prediction of college performance of ethnic minorities using the IRT-based GPA. *Applied Measurement in Education*, 4, 229–239.

Young, M., Barnes, T., Stevens, M. et al. (2007). The changing landscape of indigenous gambling in Northern Australia: Current knowledge and future directions. *International Gambling Studies,* 7, 327–343.

Zapolski, T. C. B., Pedersen, S. L., McCarthy, D. M. & Smith, G. T. (2014). Less drinking, yet more problems: Understanding African

American drinking and related problems. *Psychological Bulletin*, 140, 188-223.

Zelnik, M. & Kantner, F. M. (1980). Sexual activity, contraceptive use and pregnancy among metropolitan area teenagers. *Family Planning Perspectives*, 12, 230–238.

Zelnik, M. & Kim, Y. J. (1982). Sex education and its association with teenage sexual activity, pregnancy and contraceptive use. *Family Planning Perspectives*, 14, 117-126.

Zelnik, M. & Shah, F. K. (1983). First intercourse among young Americans. *Family Planning Perspectives*, 15, 64–70.

Zitzow, D. (1996a). Comparative study of problematic gambling behaviours between American Indian and non-Indian adolescents in a northern plains reservation. *American Aboriginal and Alaska Aboriginal Mental Health Research*, 7, 14-26.

Zitzow, D. (1996b). Comparative study of problematic gambling behaviors between American Indian and non-Indian adults in a northern plains reservation. *American Aboriginal and Alaska Aboriginal Mental Health Research*, 7, 27–41.

Zoccolillo, M., Pickles, A., Quinton, D. & Rutter, M. (1992). The outcome of childhood conduct disorder: implications for defining adult personality disorder and conduct disorder. *Psychological Medicine*, 22, 971-986.

Zubrick, S., Silburn, S., Lawrence, D. et al. (2005). *The Western Australian Aboriginal child health survey: the social and emotional wellbeing of Aboriginal children and young people.* Perth: Curtin University of Technology and Telethon Institute for Child Health Research.

Zuckerman, M. (2003). Are there racial and ethnic differences psychopathic personality? A critique of Lynn's (2002) racial and ethnic differences psychopathic personality. *Personality and Individual Differences*, 35, 1463-1469.

Zytkoskee, A., Strickland, B. R. & Watson, J. (1971). Delay of gratification and internal versus external control among adolescents of low socio-economic status. *Developmental Psychology*, 4, 93–98.

RACE DIFFERENCES IN HOMICIDE PER 100,000, EARLY TWENTY-FIRST CENTURY

COUNTRIES	HOMICIDE	COUNTRIES	HOMICIDE	COUNTRIES	HOMICIDE
EUROPE		SOUTH ASIA/ NORTH AFRICAN		NORTH EAST ASIA	
Andora	4	Armenia	4	China	1
Argentina	16	Bahrain	1	Hong Kong	1
Austria	2	Bangladesh	2	Japan	1
Belarus	10	Burma (Myanmar)	3	Korea: South	2
Belgium	13	Israel	2	Singapore	1
Bulgaria	8	Jordan	7	**Mean**	**1.2**
Chile	6	Kuwait	2		
Croatia	7	Lebanon	4	Central Asia	

COUNTRIES	HOMICIDE	COUNTRIES	HOMICIDE	COUNTRIES	HOMICIDE
Cyprus	3	Maldives	1	Kazakhstan	16
Czech Rep	3	Mauritius	3	Azerbaijan	8
Denmark	3	Nepal	2	Uzbekistan	5
Estonia	21	Oman	1	Mongolia	19
Finland	1	Qatar	2	**Mean**	12.0
France	4	Saudi Arabia	1		
Georgia	8	Sri Lanka	10	LATIN AMERICA	
Germany	5	Syria	1	Belize	24
Greece	3	Turkey	3	Colombia	66
Hungary	4	Tunisia	1	Ecuador	23
Ireland	2	UAE	1	Guyana	18
Latvia	12	**Mean**	**2.7**	Honduras	60
Lithuania	13			Panama	17
Luxemborg	1	SOUTH EAST ASIA		Paraguay	15
Macedonia	4	Brunei	2	Venezuela	23
Malta	2	Indonesia	1	Mean	30.1

APPENDICES

COUNTRIES	HOMICIDE	COUNTRIES	HOMICIDE	COUNTRIES	HOMICIDE
Moldova	10	Malaysia	2		
Netherlands	21	Thailand	8	**SUB-SAHARAN AFRICAN**	
New Zealand	3	Vietnam	2	Angola	11
Norway	3	**Mean**	**3.0**	Benin	4
Poland	3			Botswana	14
Portugal	4	Caribbean		Burkina Faso	1
Romania	5	Antigua	5	Côte d'Ivoire	3
Russia	21	Barbados	6	Ethiopia	16
Slovakia	2	Bahamas	14	Gambia	1
Slovenia	5	Bermuda	14	Ghana	2
Spain	2	Dominica	8	Madagascar	2
Sweden	10	Dominican Rep	14	Malawi	3
Switzerland	3	Grenada	8	Mauritania	1
Ukraine	9	Jamaica	31	Namibia	62
U. Kingdom	8	Saint Kitts	9	Niger	1

COUNTRIES	HOMICIDE	COUNTRIES	HOMICIDE	COUNTRIES	HOMICIDE
United States	8	Trinidad	12	South Africa	92
Mean	**6.9**	**Mean**	**13.4**	Swaziland	61
				Tanzania	8
PACIFIC ISLANDS				Uganda	9
Fiji	11			Zambia	11
Samoa	8			Zimbabwe	12
Mean	**9.5**			Mean	16.5

RACE DIFFERENCES IN CORRUPTION (CPI), 2012

COUNTRIES	CPI	COUNTRIES	CPI	COUNTRIES	CPI
EUROPE		SUB-SAHARAN AFRICAN		SOUTH ASIA/ NORTH AFRICAN	
Albania	33	Angola	22	Algeria	34
Argentina	36	Benin	36	Armenia	34
Australia	85	Botswana	65	Bahrain	51
Austria	69	Burkino Faso	38	Bangladesh	26
Belarus	31	Burundi	19	Bhutan	63
Belgium	75	Cameroon	26	Burma (Myanmar)	15
Bosnia	42	Cent African Rep	26	Egypt	32
Bulgaria	41	Chad	19	Jordan	48
Canada	84	Congo Braz	26	India	36
Chile	72	Congo Zaire	21	Iran	28
Costa Rica	59	Côte d'Ivoire	39	Iraq	18
Croatia	46	Djibouti	36	Israel	60
Cuba	48	Equatorial Guinea	30	Kuwait	44

COUNTRIES	CPI	COUNTRIES	CPI	COUNTRIES	CPI
Cyprus	66	Eritrea	25	Lebanon	30
Czech Rep	49	Ethiopia	33	Libya	21
Denmark	90	Gabon	36	Mauritius	57
Estonia	64	Gambia	34	Nepal	27
Finland	90	Ghana	45	Oman	47
France	71	Guinea	24	Pakistan	27
Georgia	52	Guinea Bissau	25	Qatar	68
Germany	79	Kenya	27	Saudi Arabia	44
Greece	36	Lesotho	45	Sri Lanka	40
Hungary	55	Liberia	41	Sudan	13
Ireland	69	Madagascar	32	Syria	26
Kosovo	34	Malawi	37	Tunisia	41
Latvia	49	Mali	34	Turkey	49
Lithuania	54	Mauritania	31	UAE	68
Luxemborg	89	Mozambique	31	Yemen	23
Macedonia	43	Namibia	48	**Mean (28)**	**38.2**
Malta	57	Niger	33		
Moldova	36	Nigeria	27	Caribbean	
Montenegro	41	Rwanda	53	Barbados	76

COUNTRIES	CPI	COUNTRIES	CPI	COUNTRIES	CPI
Netherlands	84	Senegal	36	Bahamas	71
New Zealand	90	Sierra Leone	31	Dominica	58
Norway	85	Somalia	8	Dominican Rep	32
Poland	58	South Africa	43	Haiti	19
Portugal	63	Swaziland	37	Jamaica	38
Puerto Rico	63	Tanzania	35	Saint Lucia	71
Romania	44	Togo	30	Saint Vincent	62
Russia	28	Uganda	39	Trinidad	39
Serbia	39	Zambia	37	**Mean (9)**	**52.0**
Slovakia	46	Zimbabwe	20		
Slovenia	61	**Mean (42)**	**32.4**	**CENTRAL ASIA**	
Spain	65			Azerbaijan	28
Sweden	88	**SOUTH EAST ASIA**		Kazakhstan	28
Switzerland	86	Brunei	55	Kyrgyzstan	24
Ukraine	26	Indonesia	32	Mongolia	36
U. Kingdom	74	Laos	31	Tajikistan	22
United States	73	Malaysia	49	Turkmenistan	17
Uruguay	72	Thailand	37	Uzbekistan	17

COUNTRIES	CPI	COUNTRIES	CPI	COUNTRIES	CPI
Mean (50)	63.6	Vietnam	31	**Mean (7)**	**24.7**
		Mean (6)	**39.1**		
NORTH EAST ASIA				PACIFIC ISLANDS	
China	39			Papua N.Guinea	25
Hong Kong	77			Philippines	34
Japan	74			Sao Tome	42
Korea: South	. 56			Timor Leste	33
Korea: North	8			Mean	33.5
Singapore	87				
Taiwan	61				
Mean (7)	57.4				

INDEX

A

ADHD 66
  Northeast Asians 152
  Pacific Islanders 199
  race differences 30–31
  sub-Saharan Africans 134
  White South Africans 134
Adolescent Health Survey 62, 69
Africa 4, 113, 133–149, 157, 159,
    163, 166, 176, 220, 223,
    226–228, 239, 243–244,
    248–249, 326, 329
African-Americans 3, 15–16, 19–
    21, 23–32, 34–41, 44–45,
    52–53, 60, 76, 97–99, 256
Africans. *See* African-Americans;
    See sub-Saharan Africans;
    See Black British; *See* Black
    French; *See* Black Dutch;
    See Black Canadians; *See* sub-
    Saharan Africa and Africans;
    See South Africa; *See* North
    Africa and Africans
Afro-Caribbeans 220
Afro-Latin Americans 213
aggression 35, 230, 249
  genetic basis 243–244
AIDS. *See* HIV/AIDS
AIDS Behavioral Research Project
    70
Alaska 207–211
alcoholism. *See* substance abuse
altruism 96–99, 112, 130, 249
  African-Americans 97–99

Asian-Americans 97–99
Black British 131
British East Asians 131
British South Asians 131
charitable giving 98–99
Europeans 130
Great Britain 131
Hispanics 97
Hong Kong 160
neurological basis 96
Northeast Asians 160
psychopathy, correlation with 96
sex differences 97–98
Sweden 130
United States, the 96–99, 160
American Association for Protecting
    Children 86
American General Household
    Survey 46
American National Survey of
    Adolescent Males 74
American National Survey of
    Families and Households 42
American National Survey of Family
    Growth 51
American Psychiatric Association
    4–6, 41, 58, 64, 85, 194,
    262, 263
American Psychological Association
    27, 33, 35
American Teenage Attitudes and
    Practices Survey 61
anthropology 133, 151, 163, 199
anxiety disorder. *See* mental illness
Argentina 213
Arizona 64
Asian-Americans 19, 33, 36, 45,
    58, 60, 66, 73, 82–84,
    87–91, 97–99, 240, 251
Asian Canadians 105–108
Asian New Zealanders 192, 196

Asians. *See also* Asian-Americans;
   See also Asian Canadians; *See*
   also Asian New Zealanders
Attention Deficit Hyperactivity
   Disorder. *See* ADHD
Atwood, Thomas 176
Australia 156, 179–185, 227. *See*
   also White Australians; *See*
   also Australian Aborigines
Australian Aborigines 179–187,
   223, 227
   demography 179
Ausubel, David 189

B

Bachman, J. Gerald 26
Bahamas, the 170, 176, 325, 329
Baltimore 55
Bangladesh 163
Bantus 133
Barbé, Raymond 133
Bartrusch, Dawn 32
basketball 34–35
Belgium 164
Belize 216
Bell Curve, The (Herrnstein and
   Murray) 3, 254. *See also*
   Herrnstein, Richard; *See also*
   Murray, Charles.
Bernstein, D. P. 9
birth rates
   African-Americans 73–74,
   84–85, 258
   Australia 179
   Australian Aborigines 181
   Hispanics 73–74, 84–85
   Maori, the 194
   Whites 73–74, 84–85, 194

birth weight 254, 257
   African-Americans 257
   Hispanics 257
Black British 113–120, 124–126,
   128, 130–131, 134
Black Canadians 105–106
Black Dutch 127–128
Black French 118
Blacks. *See* African-Americans;
   See Black British; *See* Black
   Canadians; *See also* Black
   Dutch; *See* Black French;
   See Caribbean Africans
blood groups 133
Blumenbach, Johann Friedrich 133,
   151
Botswana 140, 143
Brazil 213–215, 218–221, 226,
   234
British East Asians 114–118, 131
British Empire, the 147, 176
British South Asians 114–120,
   125–126, 131
bullying 117
Bureau of Justice Statistics 39
Burundi 139, 142, 327

C

California 23, 36, 72–73, 82, 88,
   92–93, 251
Cameroon 145
Canada 35, 48, 105–110, 207, 211
   Hispanics 106, 108
cancer, prostate 251
Caribbean Africans 169–176
Caribbean, the 169–176
Carothers, J.C. 133
Cartwright, Duncan 9

R

ABOUT RICHARD LYNN

Richard Lynn was born in 1930 and educated at the Bristol Grammar School and the University of Cambridge, where he graduated in Psychology in 1953 and received the Passingham Prize for the best Psychology student of the year. He also obtained his M.A. and Ph.D. at Cambridge. He has been Lecturer in Psychology at the University of Exeter (1956-1967), Professor of Psychology at the Economic and Social Research Institute, Dublin (1967-1972), and Professor and Head of the Department of Psychology at the University of Ulster (1972-1995). He was Professor Emeritus at this university until 2018, when he was stripped of this title in the wake of a media furore surrounding his research, which offended politically correct sensibilities. Lynn has been President of the Ulster Institute for Social Research since 1995.

Lynn has been a path-breaking researcher on intelligence and personality. His books include *Arousal, Attention and the Orientation Reaction* (1966), *Personality and National Character* (1972), *Dimensions of Personality (1980)*, *Educational Achievement in Japan* (1988), *Dysgenics:* (1996), *IQ and the Wealth of Nations* (2002) (jointly with Tatu Vanhanen), *Race Differences in Intelligence* (2006), *The Global Bell Curve* (2008), *Dysgenics: Second Revised Edition* (2011), *The Chosen People: A Study of Jewish Intelligence and Achievement* (2011), *IQ and Global Inequality* (2008) (with Tatu Vanhanen), *Intelligence: A Unifying Construct for the Social Sciences* (2012) (with Tatu Vanhanen), *Race and Sport: Evolution and Racial Differences in Sporting Ability* (2015) (with Edward Dutton) and *The Intelligence of Nations* (with David Becker). He has received the US Mensa Foundation Awards for Excellence in 1985, 1993, and 2006, and the Estonian Ministry of Education and Science Award for Excellence in 2010 for his work on national differences in intelligence. He is Assistant Editor of *The Mankind Quarterly*.

www.ingramcontent.com/pod-product-compliance
Lightning Source LLC
Chambersburg PA
CBHW050331270326
41926CB00016B/3411